# MASS EXTINCTIONS

## LESSONS FROM THE PAST, WARNINGS FOR THE FUTURE



MUKESH BHARTI

Made with ♥ on the Notion Press Platform
www.notionpress.com

# Contents

# Preface

The story of life on Earth is one of extraordinary resilience, beauty, and interconnection, spanning billions of years and encompassing a diversity beyond human imagination. Yet, woven into this story are moments of profound loss—mass extinctions that reshaped the world, extinguishing vast swaths of life and altering ecosystems forever. These mass extinction events have left marks on our planet's history, driving evolution in new directions and reminding us of life's fragility. Today, as scientists warn of an ongoing sixth extinction, we find ourselves facing urgent questions about the future of Earth's biodiversity and our role in protecting it.

This book, *Mass Extinctions: Lessons from The Past, Warnings for The Future*, explores each of Earth's five major extinction events and examines the critical lessons they offer. From the Ordovician-Silurian extinction that decimated marine life, to the more recent Cretaceous-Paleogene event that famously ended the age of dinosaurs, each chapter delves into the unique conditions, consequences, and evolutionary impacts of these planetary upheavals. This journey through time provides not only an understanding of what caused these extinctions but also insights into how life adapted and evolved afterward. The resilience displayed by surviving species and the resulting changes in ecosystems serve as powerful reminders of nature's capacity to regenerate even after catastrophic loss.

At the heart of this book is the pressing question of the sixth extinction. Unlike its predecessors, the current biodiversity crisis is not driven by natural events alone; human activity has accelerated the rate of species loss to levels rarely seen in the geological record. Climate change, habitat destruction, pollution, and overexploitation are rapidly transforming ecosystems and threatening countless species. As we explore the present-day context of biodiversity loss, it becomes clear that this crisis demands not only scientific understanding but also a moral reckoning with our responsibilities as stewards of the planet.

This book is written for those who seek a deeper understanding of Earth's past and an informed perspective on its future. It is my hope that by exploring the history of mass extinctions, readers will gain a sense of both urgency and possibility that we can learn from past events and take meaningful action to protect the planet's incredible biodiversity. May this journey through time inspire a commitment to conservation, sustainability, and a renewed respect for the delicate balance that sustains life on Earth.

In the pages that follow, we will explore the forces that have shaped life's history, the lessons we can draw from nature's resilience, and the steps we must take to ensure that future generations inherit a world rich with the diversity and wonder of life.

# Acknowledgements

Writing *Mass Extinctions: Lessons from The Past, Warnings for The Future* has been an illuminating journey, and it would not have been possible without the invaluable support, inspiration, and encouragement of many people to whom I owe my deepest gratitude.

First and foremost, I would like to extend my sincere thanks to the scientists and researchers whose tireless work in the fields of paleontology, geology, ecology, and environmental science has made this exploration of Earth's mass extinctions both fascinating and enlightening. Their dedication to uncovering the secrets of the past and their commitment to conservation have shaped the foundation of this book.

I am profoundly grateful to my mentors and colleagues, whose expertise and guidance provided me with the insights necessary to bring this complex story to life. Their constructive feedback and encouragement have helped refine my understanding of mass extinctions, the ongoing biodiversity crisis, and the essential lessons we can draw from our planet's history.

To my friends and family, thank you for your unwavering support, patience, and encouragement throughout this endeavor. Your belief in my vision and your enthusiasm for this project have been a constant source of motivation. I am especially grateful for your understanding as I immersed myself in the history of life on Earth, spending countless hours reading, researching, and writing.

Special thanks are due to the publishing team, whose hard work and expertise brought this book to life. From editorial guidance to design, they have helped transform my manuscript into a polished work that I am proud to share with readers. Their commitment to making this complex subject accessible and engaging is deeply appreciated.

Finally, I would like to thank you, the reader. Your curiosity and passion for understanding our world's history and future make this work meaningful. I hope this book inspires a deeper appreciation for the richness of life on Earth and a commitment to protecting it for generations to come.

**With gratitude,**

***Mukesh Bharti***

# Purpose Of The Book

The purpose of *Mass Extinctions: Lessons from The Past, Warnings for The Future* is to take readers on an in-depth journey through the history of Earth's mass extinctions, examining each of the five major extinction events and culminating in an analysis of the current biodiversity crisis often referred to as the "sixth extinction." Through this exploration, the book aims to shed light on both the fragility and resilience of life on our planet, using each extinction event as a lens to understand the powerful forces that have shaped Earth's biological and ecological history.

By understanding the causes, impacts, and aftermaths of past extinctions, this book seeks to deepen our appreciation for the complexities of ecosystems and the delicate balance required to sustain life. Each chapter is designed to provide readers with a detailed account of these events, not only from a scientific perspective but also through a broader lens that considers the philosophical, ethical, and environmental implications of mass extinctions.

Another core purpose of this book is to bridge the gap between scientific knowledge and urgent social responsibility. The current biodiversity crisis, largely driven by human activity, is unlike any previous extinction in both its speed and scope. This book strives to highlight the role of human impact—climate change, habitat destruction, pollution, and overexploitation of resources—and to foster a sense of collective accountability, encouraging readers to engage in conservation efforts and adopt sustainable practices.

Ultimately, *Mass Extinctions: Lessons from The Past, Warnings for The Future* aims to inspire a transformation in our relationship with the natural world. Through an understanding of the past, we can better grasp the stakes of the present and take informed action to protect our planet's future. The book calls on readers to reflect on humanity's role as both caretakers and disruptors, urging each of us to make choices that will contribute to a legacy of preservation rather than depletion. This book, therefore, serves as both a history of life's tenacity and a call to action, reminding us that the story of Earth's biodiversity is one we have the power to influence.

CHAPTER ONE

# INTRODUCTION

## The Cycle of Life and Death on Earth

The history of life on Earth is a story of creation, adaptation, and extinction. For over 4.5 billion years, our planet has seen the rise and fall of countless species. Among the myriad changes that have occurred throughout Earth's geological timeline, few events have had as profound an impact on the course of life as mass extinctions. These cataclysmic occurrences, often caused by a combination of environmental changes and astronomical or geological phenomena, have wiped out large percentages of Earth's species, fundamentally altering the trajectory of evolution.

Mass extinctions are defined as periods in Earth's history when a significant proportion of all living species disappeared in a relatively short geological time frame. These events typically result in the loss of at least 75% of species across a wide range of habitats, both marine and terrestrial. Unlike background extinctions, which occur gradually over time as species evolve and adapt to changing environments, mass extinctions are abrupt, catastrophic, and often global in scale.

The study of mass extinctions gives us critical insights into the fragility of life and the delicate balance required for ecosystems to thrive. These events force us to confront the reality that life on Earth is not guaranteed permanence, and that species, no matter how dominant or well-adapted, can be brought to the brink of annihilation by sudden environmental shifts.

In this book, we will explore in detail the five major mass extinctions that have reshaped life on Earth. From the devastating end of the Permian Period, where nearly all life on the planet was extinguished, to the famous demise of the dinosaurs during the Cretaceous-Paleogene extinction, each of these events serves as a stark reminder of the unpredictable nature of life on Earth. But perhaps the most alarming aspect of this exploration is the understanding that we are currently living through the sixth mass extinction, an event unlike any other in Earth's history due to its primary cause: human activity.

**The Nature of Extinction**

To truly grasp the gravity of mass extinctions, we must first understand the broader concept of extinction itself. Extinction is a natural process. Over 99% of all species that have ever existed on Earth are now extinct. Extinctions happen for a multitude of reasons: species fail to adapt to new predators, their habitats disappear, or they fall victim to diseases. Most extinctions occur slowly, over thousands or millions of years, as species are outcompeted by others that are better adapted to the environment.

However, the relatively slow pace of normal, background extinctions is drastically different from what occurs during a mass extinction. A mass extinction is a rapid, wholesale annihilation of species across the globe. It happens in a geologically short period of time, often spanning less than a few million years. Such an event marks a dramatic reduction in biodiversity, and ecosystems are forced to rebuild from the ground up.

Mass extinctions are critical junctures in Earth's history, acting as evolutionary bottlenecks. These events eradicate species that were once thriving, leveling the playing field for those that survive. In the aftermath, life diversifies and fills the empty ecological niches left behind, resulting in the rise of new species and the reshaping of ecosystems. This process, known as adaptive radiation, has led to some of the most remarkable evolutionary developments in Earth's history.

**The Importance of Studying Mass Extinctions**

The study of past mass extinctions offers us a window into Earth's biological and geological history, providing vital information on how life has responded to periods of extreme stress. By examining the causes and consequences of these events, scientists can better understand the forces that shape biodiversity and the complex interactions between life and the environment.

Mass extinctions are not isolated events; they are deeply interconnected with the processes of plate tectonics, climate change, ocean circulation, and atmospheric composition. The causes of these extinctions are often multifaceted, involving a combination of volcanic eruptions, asteroid impacts, sea-level changes, and shifts in the Earth's climate. Understanding how these factors contributed to the collapse of ecosystems in the past helps us predict how modern ecosystems might respond to current environmental pressures.

In particular, the study of mass extinctions has become more urgent as scientists recognize the parallels between past events and the changes currently affecting our planet. Climate change, habitat destruction, pollution, and overexploitation of resources are driving unprecedented rates of species loss, leading many experts to conclude that we are in the midst of the sixth mass extinction. Unlike previous extinctions, this one is being caused primarily by human activity, making it a unique and troubling chapter in Earth's history.

**The Role of Geology and Paleontology**

Geologists and paleontologists play a crucial role in unraveling the mysteries of mass extinctions. Through the study of rock layers, fossil records, and isotopic analysis, these scientists piece together the timing, magnitude, and causes of extinction events. The fossil record, in particular, serves as a time capsule, preserving the remains of ancient life forms and allowing us to trace the rise and fall of species over millions of years.

The geological evidence for mass extinctions is often found in the form of boundary layers—thin bands of sediment that mark the transition between different geological periods. These layers contain the remnants of organisms that perished during the extinction event, as well as chemical signatures that hint at the environmental conditions at the time. For example, the Cretaceous-Paleogene boundary layer, which marks the end of the age of dinosaurs, contains high levels of iridium, an element rare on Earth but abundant in asteroids. This discovery provided crucial evidence for the asteroid impact hypothesis, which is now widely accepted as the primary cause of the extinction.

In addition to boundary layers, paleontologists study the distribution of fossils before, during, and after mass extinctions to understand how ecosystems were affected. By comparing the types of species that survived with those that perished, scientists can identify patterns of vulnerability and resilience. For example, many mass extinctions disproportionately affected large, specialized species, while smaller, more adaptable organisms often fared better. These insights help us understand the factors that influence a species' ability to survive environmental stress.

**The Five Major Mass Extinctions**

Throughout Earth's history, five mass extinctions stand out as particularly significant in their scope and impact. These events, often referred to as the "Big Five," have shaped the course of evolution and left indelible marks on the planet's geological record.

**The Ordovician-Silurian Extinction (443 million years ago)**: This was one of the earliest mass extinctions, and it primarily affected marine life. At the time, most life on Earth was confined to the oceans, and this event wiped out approximately 85% of marine species. It was likely caused by a combination of a severe ice age, falling sea levels, and changes in ocean chemistry. The Ordovician-Silurian extinction marks a critical point in the evolution of marine ecosystems, as new species emerged to fill the ecological gaps left behind.

**The Late Devonian Extinction (375-360 million years ago)**: This extinction event unfolded over a prolonged period, spanning millions of years. It affected both marine and terrestrial ecosystems, with a particular impact on coral reefs and armored fish. The causes of the Late Devonian extinction are still debated, but they likely include climate cooling, volcanic activity, and changes in ocean circulation. This extinction paved the way for the rise of land-based ecosystems, as plants and early vertebrates began to dominate terrestrial environments.

**The Permian-Triassic Extinction (252 million years ago)**: Often referred to as "The Great Dying," the Permian-Triassic extinction is the most severe mass extinction in Earth's history. Approximately 96% of all marine species and 70% of terrestrial species were wiped out. This event was likely triggered by massive volcanic eruptions in what is now Siberia, leading to a cascade of environmental changes, including global warming, ocean acidification, and anoxic conditions in the oceans. The Permian-Triassic extinction marks a turning point in Earth's history, as the dominance of early reptiles gave way to the rise of the dinosaurs.

**The Triassic-Jurassic Extinction (201 million years ago)**: This extinction event cleared the way for the age of dinosaurs. It was likely caused by volcanic activity associated with the breakup of the supercontinent Pangaea, leading to climate change and ocean acidification. The extinction affected a wide range of species, including large amphibians and early reptiles, but it also opened up ecological niches that allowed dinosaurs to thrive for the next 135 million years.

**The Cretaceous-Paleogene (K-Pg) Extinction (66 million years ago)**: The most famous of the mass extinctions, the K-Pg extinction marks the end of the age of dinosaurs. It was caused by a combination of factors, including a massive asteroid impact and volcanic activity. The resulting environmental changes, including wildfires, acid rain, and global cooling, led to the extinction of approximately 75% of all species, including all non-avian dinosaurs. This event set the stage for the rise of mammals and, eventually, the evolution of humans.

**The Sixth Mass Extinction: A Human-Made Crisis**

Today, scientists warn that we are in the midst of a sixth mass extinction, driven primarily by human activities. Unlike previous extinctions, which were caused by natural events such as volcanic eruptions or asteroid impacts, the current crisis is the result of habitat destruction, climate change, pollution, and overexploitation of resources. Species are disappearing at an alarming rate, with some estimates suggesting that current extinction rates are 100 to 1,000 times higher than the natural background rate.

The sixth mass extinction poses unique challenges because it is occurring during a time when human civilization is thriving. The loss of biodiversity not only threatens ecosystems but also undermines the resources and services that humans depend on for survival. Ecosystems provide critical functions such as pollination, water purification, and climate regulation, all of which are at risk as species disappear.

However, there is still hope. Unlike previous mass extinctions, we have the knowledge and the tools to mitigate the damage. Conservation efforts, habitat restoration, and sustainable practices can help slow the rate of species loss and protect the planet's biodiversity for future generations. The key lies in recognizing the urgency of the situation and taking collective action to address the root causes of the sixth extinction.

**Conclusion**

The study of mass extinctions offers profound insights into the history of life on Earth. These events, while devastating, have also paved the way for new forms of life to emerge and thrive. As we face the challenges of the sixth mass extinction, it is crucial to remember the lessons of the past and take proactive steps to preserve the rich diversity of life that makes our planet unique. In the pages that follow, we will explore each of the five major mass extinctions in detail, culminating in an in-depth analysis of the current extinction crisis and what can be done to prevent further loss of biodiversity.

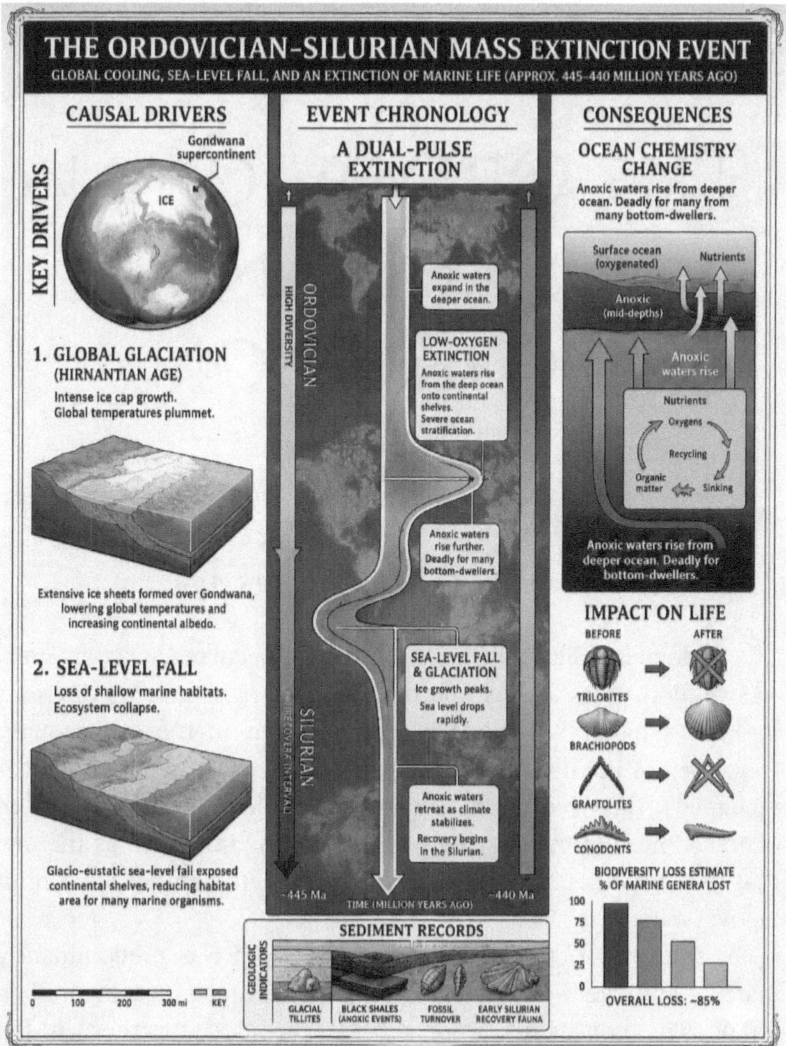


FIGURE 1.1: CHRONOLOGY AND IMPACTS OF THE LATE ORDOVICIAN MASS EXTINCTION.

Ordovician & Silurian Mass Extinction

CHAPTER TWO

# THE ORDOVICIAN-SILURIAN EXTINCTION

### *Around 443 Million Years Ago*

The Ordovician-Silurian extinction, which occurred approximately 443 million years ago, is one of the five largest mass extinctions in Earth's history. It marks a critical point in the evolutionary trajectory of life on our planet, and its far-reaching consequences reshaped the biosphere. This extinction event is often overshadowed by more widely known extinctions such as the one that led to the demise of the dinosaurs, but its significance cannot be overstated.

At the time of the Ordovician period, life was predominantly marine, and the world was a vastly different place compared to today. The continents were mostly clustered together in the Southern Hemisphere, forming a supercontinent known as Gondwana. The climate was warm, and life in the oceans was diverse and thriving. Yet, within a relatively short span of time, the Earth underwent dramatic environmental changes that wiped out nearly 85% of all marine species, making the Ordovician-Silurian extinction one of the most catastrophic biological crises the planet had ever experienced.

**The Ordovician World: A Flourishing Marine Ecosystem**

To fully appreciate the impact of the Ordovician-Silurian extinction, it's important to understand what life was like during the Ordovician period, which lasted from around 485 million to 443 million years ago. During this time, the Earth was dominated by marine ecosystems. Life had not yet colonized land in any significant way, and the oceans were teeming with a diverse array of organisms, many of which would be unrecognizable today.

The Ordovician period is often described as a time of evolutionary experimentation. The oceans were filled with a wide variety of invertebrates, including trilobites, brachiopods, bryozoans, mollusks, and the first coral reefs. Graptolites, a group of colonial animals, were especially abundant and serve as key index fossils for this period. Fish, the earliest vertebrates, were beginning to diversify, though they were still primarily jawless.

During the Ordovician, marine life flourished in warm, shallow seas that covered much of the Earth. The supercontinent Gondwana, which included modern-day Africa, South America, Australia, Antarctica, and parts of Asia, was located near the South Pole, but large portions of it were submerged under shallow seas. These seas were ideal habitats for a wide range of organisms, many of which were sessile, meaning they were anchored to the sea floor. Filter-feeding organisms such as brachiopods and bryozoans dominated these environments, and complex reef systems provided shelter for a variety of other species.

One of the most remarkable aspects of the Ordovician period was the diversification of marine life in what is often referred to as the "Ordovician Radiation" or the "Great Ordovician Biodiversification Event." This was a period of rapid evolutionary expansion, during which the number of marine genera and species increased dramatically. New ecological niches were created, and marine ecosystems became more complex and structured. The biodiversity of the Ordovician oceans rivaled that of modern marine ecosystems, with an impressive variety of species occupying different ecological roles.

The Ordovician period also saw significant changes in Earth's climate and geography. The supercontinent Gondwana was gradually drifting toward the South Pole, and sea levels were relatively high. The warm climate supported extensive shallow marine environments, which were ideal for the proliferation of life. However, this stable and thriving world was about to face a dramatic upheaval.

**The Causes of the Ordovician-Silurian Extinction**

The Ordovician-Silurian extinction occurred in two main pulses, with the first wave of extinctions happening at the end of the Ordovician period and a second wave during the early Silurian period. The causes of this extinction event are complex and multifaceted, but most scientists agree that it was triggered by a combination of global cooling, glaciation, and changes in sea levels.

One of the primary drivers of the Ordovician-Silurian extinction was a sudden and severe ice age. As the supercontinent Gondwana moved closer to the South Pole, the climate began to cool. This cooling trend eventually led to the formation of extensive ice sheets that covered much of the Southern Hemisphere. The rapid onset of glaciation had a profound impact on global climate, sea levels, and ocean chemistry, all of which contributed to the mass extinction.

**Glaciation and Climate Change**

The glaciation that occurred at the end of the Ordovician was one of the most significant climate events in Earth's history. As ice sheets grew and expanded across Gondwana, global temperatures dropped, and the Earth's climate shifted from a warm, greenhouse state to a cooler, icehouse state. This cooling was likely exacerbated by changes in atmospheric carbon dioxide levels. Some studies suggest that a reduction in volcanic activity during this time may have led to a decrease in the amount of CO2 being released into the atmosphere, which contributed to the cooling trend.

The onset of glaciation had a dramatic effect on sea levels. As more and more water was locked up in ice sheets, sea levels began to fall, exposing vast areas of the continental shelf that had previously been submerged under shallow seas. These shallow marine environments were home to many of the species that would be affected by the extinction, including coral reefs, brachiopods, and trilobites. As sea levels dropped, these habitats were destroyed, leading to a collapse in biodiversity.

The cooling of the climate also affected ocean circulation patterns and the chemistry of the oceans. As the Earth cooled, the temperature gradient between the equator and the poles increased, leading to changes in ocean currents. These changes likely disrupted the distribution of nutrients in the oceans, which would have had a cascading effect on marine ecosystems. The cooling of the oceans also made it more difficult for some species to survive, particularly those that were adapted to warmer waters.

**Sea-Level Changes and Habitat Loss**

The dramatic changes in sea level during the Ordovician-Silurian extinction had a profound impact on marine life. As the ice sheets grew and sea levels fell, many shallow marine habitats were destroyed. These habitats, which included reefs, lagoons, and coastal areas, were home to a wide variety of species that were highly specialized and dependent on stable environmental conditions. The loss of these habitats led to a decline in biodiversity, as many species were unable to adapt to the rapidly changing conditions.

In addition to the direct effects of habitat loss, the falling sea levels also contributed to changes in ocean chemistry. As shallow seas receded, they exposed large areas of the seafloor, which altered the balance of nutrients and dissolved gases in the oceans. This led to a reduction in oxygen levels, particularly in deeper waters, creating anoxic conditions that were lethal to many marine organisms. The combination of habitat loss, changes in ocean circulation, and declining oxygen levels created a perfect storm for mass extinction.

**Ocean Chemistry and Anoxia**

Another critical factor in the Ordovician-Silurian extinction was the development of anoxic conditions in the oceans. Anoxia occurs when oxygen levels in the water drop to dangerously low levels, creating a hostile environment for most marine life. During the late Ordovician, the cooling climate and changes in ocean circulation patterns led to widespread anoxia in the deep oceans.

The development of anoxic conditions is closely tied to the disruption of ocean circulation. In a stable climate, warm water at the surface of the ocean is continually mixed with cooler, deeper water, which helps to distribute oxygen and nutrients throughout the water column. However, during periods of rapid cooling, this circulation can be disrupted, leading to a stratified ocean where the deeper waters become isolated from the surface and lose their oxygen supply.

The anoxic conditions that developed during the Ordovician-Silurian extinction were particularly devastating for marine life. Many species, particularly those that lived in deeper waters or were less mobile, were unable to survive in the low-oxygen environments. The collapse of these ecosystems had a cascading effect, as the loss of key species disrupted the entire food web.

**Volcanic Activity and Carbon Dioxide Levels**

In addition to glaciation and changes in sea levels, there is evidence to suggest that volcanic activity may have played a role in the Ordovician-Silurian extinction. Volcanic eruptions can have a significant impact on the Earth's climate by releasing large amounts of gases, such as carbon dioxide ($CO2$) and sulfur dioxide ($SO2$), into the atmosphere. These gases can influence global temperatures and ocean chemistry.

During the late Ordovician, there is evidence of widespread volcanic activity, particularly in what is now North Africa. Some scientists believe that this volcanic activity may have contributed to the cooling of the climate by releasing sulfur dioxide into the atmosphere, which would have reflected sunlight and lowered global temperatures. At the same time, volcanic eruptions release large amounts of carbon dioxide, a greenhouse gas that can warm the planet. The interplay between these two effects; cooling from sulfur dioxide and warming from carbon dioxide, may have contributed to the climate instability that characterized the late Ordovician.

In addition to its impact on climate, volcanic activity can also affect ocean chemistry. The release of volcanic gases and ash into the atmosphere can lead to acid rain, which can lower the pH of the oceans and disrupt marine ecosystems. Acidification of the oceans may have been another factor that contributed to the Ordovician-Silurian extinction, particularly in shallow marine environments where the effects of acid rain would have been most pronounced.

**The Impact of the Ordovician-Silurian Extinction**

The Ordovician-Silurian extinction was a global event that affected marine ecosystems across the planet. It is estimated that approximately 85% of all marine species were wiped out, making it one of the most severe mass extinctions in Earth's history. The extinction had a particularly devastating impact on shallow marine environments, where the majority of Ordovician life was concentrated.

Among the hardest-hit groups were the trilobites, a class of arthropods that had been one of the dominant forms of life in the oceans for hundreds of millions of years. While trilobites had already experienced some decline prior to the extinction, the Ordovician-Silurian event dealt them a severe blow, and they never fully recovered. Many species of brachiopods, bryozoans, and corals also went extinct during this time.

The extinction also had a significant impact on the biodiversity of marine ecosystems. The Ordovician period had been a time of rapid diversification, with new species evolving to fill a variety of ecological niches. The extinction wiped out many of these newly evolved species, reducing the complexity of marine ecosystems. However, in the aftermath of the extinction, new groups of organisms began to rise to prominence, setting the stage for the next phase of evolutionary history.

**Recovery and the Silurian Period**

In the wake of the Ordovician-Silurian extinction, the Earth entered a period of recovery and renewal. The Silurian period, which followed the extinction, saw the gradual restoration of marine ecosystems and the emergence of new forms of life. While the extinction had dramatically reduced biodiversity, it also created new opportunities for the surviving species to evolve and diversify.

One of the most significant developments during the Silurian period was the rise of jawed fish, which would go on to become one of the dominant groups of vertebrates in the oceans. The recovery of coral reefs and other marine habitats also allowed for the reestablishment of complex ecosystems. However, the Silurian period was not without its challenges. The climate remained unstable, and there were several smaller extinction events that punctuated the recovery process.

The Ordovician-Silurian extinction serves as a reminder of the fragility of life on Earth and the complex interplay between climate, geology, and biology. While the causes of the extinction were largely natural, the lessons it offers are still relevant today as we face the ongoing challenges of climate change and biodiversity loss. By studying the past, we can gain valuable insights into the factors that drive mass extinctions and take steps to prevent similar catastrophes in the future.

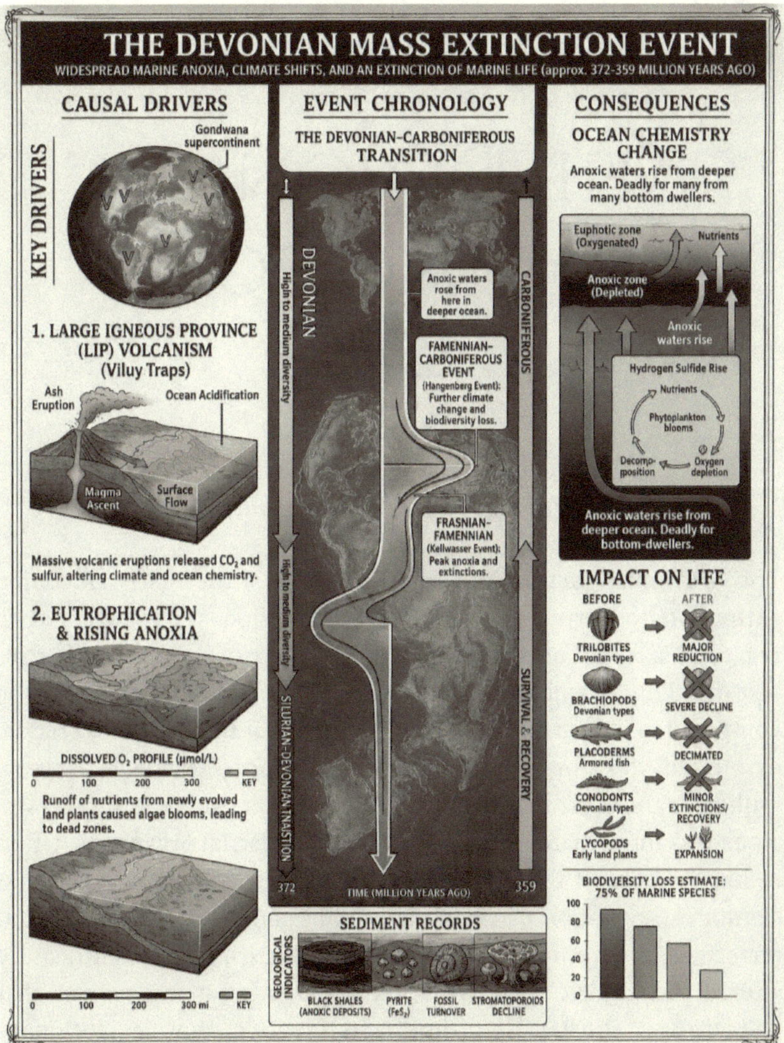

FIGURE 2.1: CHRONOLOGY AND IMPACTS OF THE LATE DEVONIAN MASS EXTINCTION.

The Devonian Mass Extinction

CHAPTER THREE

# The Devonian Extinction

### *Around 375-360 Million Years Ago*

The Devonian Extinction, also known as the Late Devonian extinction, occurred around 375-360 million years ago and is regarded as one of the five major mass extinctions in Earth's history. This extinction event, while less sudden and dramatic compared to others like the end-Cretaceous or the Permian-Triassic extinction, was a prolonged and complex process that lasted several million years. It profoundly affected marine life and had significant, albeit less immediate, consequences for terrestrial organisms. Often called the "Age of Fishes," the Devonian period witnessed the first significant evolution of vertebrates and plants, both in marine and terrestrial environments. The Devonian extinction significantly altered the trajectory of evolution, eliminating many species that had thrived during this time and creating new evolutionary opportunities for others.

To understand the Devonian extinction and its causes, one must explore the complex interplay between environmental changes, biological factors, and geological processes. The extinction's causes are multifaceted, involving both marine and terrestrial factors, including ocean anoxia (oxygen depletion), fluctuations in sea levels, changes in climate, asteroid impacts, volcanic activity, and the spread of terrestrial plants.

**The Devonian World: A Time of Evolutionary Innovation**

The Devonian period, which spanned from about 419 to 359 million years ago, was a time of remarkable evolutionary progress. This period is often remembered as the "Age of Fishes" due to the enormous diversification and dominance of fish in marine ecosystems. Fish were not the only important players, however; the Devonian period also saw the colonization of land by plants and early tetrapods, setting the stage for the development of complex terrestrial ecosystems.

The oceans of the Devonian period were filled with a variety of fish species, many of which were armored and bore little resemblance to the fish we see today. Placoderms, a class of armored fish, were among the dominant predators of the time, with some species growing to impressive sizes. Among these placoderms was the fearsome *Dunkleosteus*, a giant predator that could reach lengths of up to 10 meters and had incredibly powerful jaws. Jawless fish, such as the agnathans, were also present, but they were gradually being replaced by more advanced forms of fish with jaws, including the ancestors of modern sharks and bony fish.

In addition to the vast diversity of fish, the Devonian seas were also home to numerous invertebrate species, including brachiopods, trilobites, ammonites, and corals. Reef systems were particularly extensive during this period, with large, biologically complex reefs forming in warm, shallow seas. These reefs were built primarily by stromatoporoids (a type of sponge) and tabulate corals, which formed massive reef complexes that were home to a variety of marine organisms.

While life in the oceans flourished, the Devonian period also marked a crucial turning point in the history of life on land. Plants, which had previously been restricted to aquatic environments, began to colonize terrestrial habitats in earnest during the Devonian. Early vascular plants, such as *Rhynia* and *Psilophyton*, evolved to thrive in dry environments and helped to transform the Earth's landscapes. By the end of the Devonian, the first true forests had appeared, composed of towering, tree-like plants such as *Archaeopteris*. These plants played a vital role in stabilizing the Earth's soils, regulating the climate, and providing habitats for early terrestrial animals, including the first amphibians and arthropods.

The diversification of plants had profound effects on the Earth's atmosphere and climate. As plants spread across the continents, they absorbed significant amounts of carbon dioxide (CO2) from the atmosphere through photosynthesis, potentially contributing to global cooling. In addition, the decay of plant material released organic carbon into the soil and oceans, influencing ocean chemistry and possibly triggering some of the environmental changes that would contribute to the Devonian extinction.

**The Nature of the Devonian Extinction**

Unlike other mass extinctions that occurred suddenly and over short geological timescales, the Devonian extinction was a prolonged event that unfolded over millions of years. It is considered to be a series of extinction pulses rather than a single catastrophic event. These pulses affected marine life most severely, leading to the extinction of an estimated 75% of all marine species. Terrestrial organisms, while impacted, were not as drastically affected as their marine counterparts.

The Devonian extinction can be broken down into several distinct phases, with the most significant extinction events occurring during the late Frasnian and Famennian stages of the Devonian. These two stages were marked by rapid and widespread losses of biodiversity, particularly in marine environments. The exact causes of the extinction are still debated among scientists, but it is generally accepted that a combination of factors, both biotic and abiotic, contributed to the decline of Devonian ecosystems.

**The Role of Ocean Anoxia**

One of the most widely accepted explanations for the Devonian extinction is the development of widespread anoxia in the Earth's oceans. Anoxia refers to a condition in which the oxygen levels in the water are severely depleted, making it impossible for most marine life to survive. During the Devonian period, large portions of the ocean became anoxic, particularly in deeper waters, leading to the collapse of marine ecosystems.

The causes of ocean anoxia during the Devonian extinction are complex and likely involve multiple factors. One possible explanation is that the spread of terrestrial plants contributed to the depletion of oxygen in the oceans. As plants colonized the land, they absorbed large amounts of carbon dioxide from the atmosphere, potentially leading to a decrease in global temperatures and changes in ocean circulation patterns. This, in turn, may have disrupted the normal mixing of ocean waters, preventing oxygen from being transported from the surface to the deeper layers of the ocean.

Another factor that may have contributed to ocean anoxia is the increased weathering of rocks on land, which was facilitated by the spread of plants. As plants spread across the continents, they released organic acids that broke down rocks and minerals, releasing nutrients such as phosphorus and nitrogen into the oceans. This influx of nutrients could have triggered massive algal blooms, which, when they died and decomposed, consumed large amounts of oxygen, leading to anoxic conditions.

The anoxia hypothesis is supported by the presence of black shales in the geological record from the late Devonian. Black shales are sedimentary rocks that form in oxygen-poor environments and are often associated with mass extinction events. These rocks contain high levels of organic material, indicating that large amounts of dead organisms accumulated on the seafloor, where they could not decompose properly due to the lack of oxygen. The widespread occurrence of black shales in late Devonian deposits suggests that anoxia played a significant role in the extinction event.

**Climate Change and Glaciation**

In addition to ocean anoxia, climate change is believed to have been a major driver of the Devonian extinction. During the early and middle Devonian, the Earth experienced a relatively warm climate, with high levels of atmospheric carbon dioxide. However, as the Devonian progressed, the climate began to cool, possibly as a result of the spread of land plants and the associated reduction in atmospheric CO2.

The cooling of the climate may have contributed to the extinction by altering ocean circulation patterns and creating conditions that favored the development of anoxia. In addition, there is evidence to suggest that glaciation occurred during the late Devonian, particularly in the southern hemisphere, where the supercontinent Gondwana was located. Glaciation would have caused fluctuations in sea levels, which could have disrupted shallow marine habitats and contributed to the extinction of many species.

One of the most important pieces of evidence for late Devonian glaciation comes from geological deposits in what is now South America. These deposits contain glacial till, a type of sedimentary rock that forms from the debris left behind by glaciers. The presence of glacial till in late Devonian rocks indicates that ice sheets were present in the southern hemisphere during this time, suggesting that global cooling played a role in the extinction event.

**The Spread of Terrestrial Plants and Ecosystem Disruption**

The Devonian period saw the rapid spread of vascular plants across the land, a development that had profound effects on both terrestrial and marine ecosystems. While the colonization of land by plants was a major evolutionary milestone, it also contributed to the environmental changes that led to the Devonian extinction.

As plants spread across the continents, they altered the Earth's atmosphere and climate. Through photosynthesis, plants absorbed large amounts of carbon dioxide from the atmosphere, which may have contributed to the global cooling observed during the late Devonian. In addition, the development of root systems allowed plants to break down rocks and release nutrients into the soil, which were then washed into rivers and oceans. This influx of nutrients likely contributed to the development of algal blooms and the subsequent depletion of oxygen in marine environments.

The spread of plants also had significant impacts on terrestrial ecosystems. The first forests appeared during the late Devonian, composed of tall, tree-like plants such as *Archaeopteris*. These forests provided new habitats for a variety of organisms, including early terrestrial vertebrates and arthropods. However, the development of forests also contributed to soil erosion, as the roots of plants broke down rocks and released sediment into rivers and oceans. This increased sedimentation may have further disrupted marine ecosystems by smothering coral reefs and other habitats.

The Devonian period also saw the evolution of the first true soils, which played a crucial role in regulating the Earth's carbon cycle. As plants decayed, they released organic carbon into the soil, where it could be stored for long periods of time. This process helped to regulate atmospheric carbon dioxide levels and may have contributed to the global cooling observed during the late Devonian.

While the spread of plants had many positive effects on the Earth's ecosystems, it also created new environmental challenges that contributed to the Devonian extinction. The changes in the carbon cycle, the increased weathering of rocks, and the alteration of marine and terrestrial habitats all played a role in the environmental disruptions that led to the extinction.

**Asteroid Impact Hypothesis**

Another possible cause of the Devonian extinction is the impact of an asteroid or comet. While there is less evidence for an impact during the Devonian than there is for other extinction events, such as the end-Cretaceous extinction, some scientists have suggested that an impact could have played a role in triggering the environmental changes that led to the extinction.

The evidence for an asteroid impact during the Devonian is based primarily on the discovery of a layer of iridium-rich sediment in late Devonian rocks. Iridium is a rare element on Earth but is more common in asteroids and comets. The presence of iridium in the geological record has been used as evidence for asteroid impacts, most famously in the case of the end-Cretaceous extinction. However, the evidence for an impact during the Devonian is not as conclusive, and other factors, such as volcanic activity or changes in ocean chemistry, could also explain the presence of iridium.

If an impact did occur during the Devonian, it could have contributed to the extinction by causing widespread environmental disruption. An impact would have released massive amounts of energy, potentially triggering wildfires, tsunamis, and changes in climate. The release of dust and debris into the atmosphere could have blocked out sunlight, leading to a temporary cooling of the Earth's surface and disrupting photosynthesis in both marine and terrestrial ecosystems.

**Volcanism and the Release of Greenhouse Gases**

Volcanic activity is another possible contributor to the Devonian extinction. While there is less evidence for large-scale volcanism during the late Devonian compared to other extinction events, such as the Permian-Triassic extinction, some scientists have suggested that volcanic eruptions could have played a role in triggering environmental changes.

Volcanic eruptions release large amounts of carbon dioxide and other greenhouse gases into the atmosphere, which can lead to global warming. In addition, volcanic eruptions can release sulfur dioxide, which can combine with water vapor in the atmosphere to form sulfuric acid, leading to acid rain. Acid rain can have devastating effects on both terrestrial and marine ecosystems, particularly by disrupting the chemistry of freshwater and marine environments.

If volcanic activity did occur during the late Devonian, it could have contributed to the extinction by altering the Earth's climate and ocean chemistry. However, the evidence for large-scale volcanism during this time is limited, and it is likely that other factors, such as ocean anoxia and climate change, played a more significant role in the extinction.

**The Impact on Marine and Terrestrial Life**

The Devonian extinction had a profound impact on marine ecosystems, particularly on reef-building organisms and the diverse array of fish species that dominated the oceans during this time. The extinction wiped out many of the dominant marine species, including the placoderms, which had been the top predators of the Devonian seas. Other groups, such as the brachiopods and trilobites, also suffered significant losses, although they did not go extinct entirely.

Reef ecosystems were particularly hard hit by the extinction. The extensive reef systems that had flourished during the early and middle Devonian were decimated by the environmental changes that occurred during the late Devonian. The collapse of these reefs had far-reaching consequences for marine biodiversity, as reefs provide important habitats for a variety of organisms. It would take millions of years for reef ecosystems to recover fully, and they would never reach the same level of complexity as they had during the Devonian.

Terrestrial ecosystems were also affected by the Devonian extinction, although the impact was less severe than in the oceans. The spread of plants and the development of forests had transformed the Earth's landscapes by the end of the Devonian, creating new habitats for a variety of organisms. However, the environmental changes that contributed to the extinction, such as climate change and changes in the carbon cycle, also had consequences for terrestrial life.

One of the most significant developments during the Devonian was the evolution of the first tetrapods, or four-limbed vertebrates. These early tetrapods were the ancestors of modern amphibians, reptiles, and mammals, and their emergence during the Devonian set the stage for the evolution of complex terrestrial ecosystems in the following periods. However, the environmental disruptions of the late Devonian likely made it difficult for these early tetrapods to thrive, and many species may have gone extinct as a result of the extinction event.

**The Legacy of the Devonian Extinction**

The Devonian extinction was a turning point in the history of life on Earth. While it was not as sudden or dramatic as some of the other mass extinctions, its effects were far-reaching and long-lasting. The extinction eliminated many of the dominant species of the Devonian, including the placoderms and the reef-building organisms, and created new opportunities for the evolution of other groups.

In the aftermath of the extinction, new groups of fish, such as the ancestors of modern sharks and bony fish, began to rise to prominence in the oceans. On land, the evolution of tetrapods and the spread of forests would pave the way for the development of complex terrestrial ecosystems in the following periods. The Devonian extinction serves as a reminder of the complex interplay between environmental changes and the evolution of life, and the importance of studying past extinction events to understand the challenges we face today.

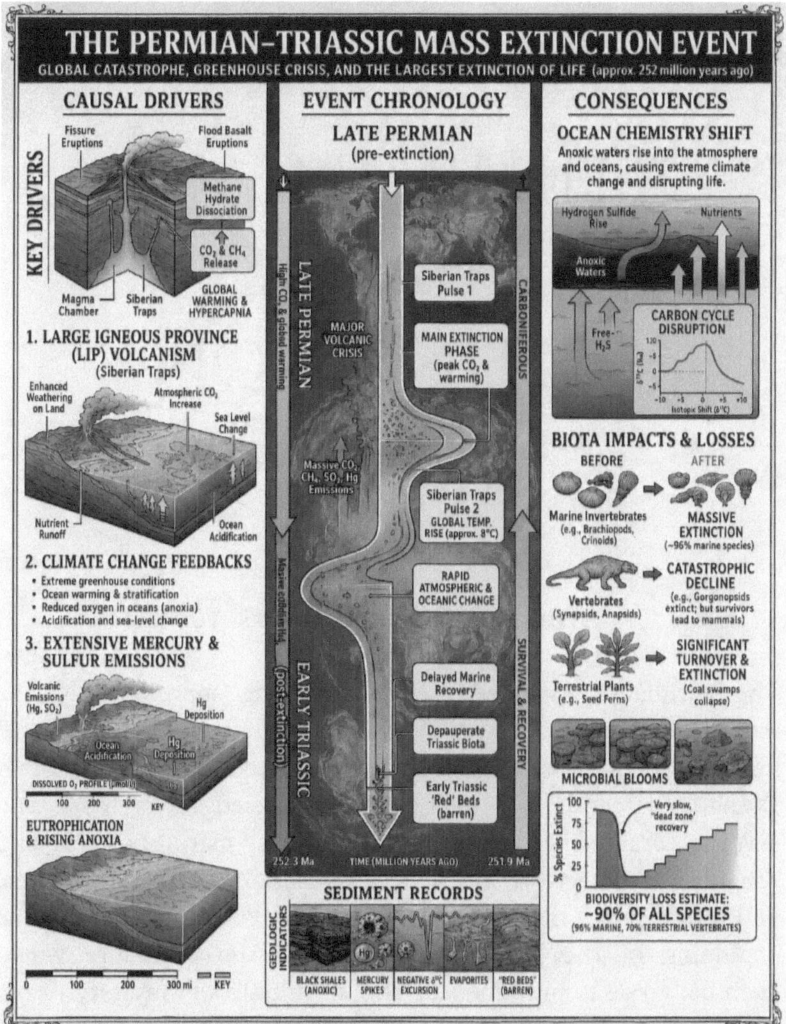


FIGURE 3.1: CAUSES, CHRONOLOGY, AND CATASTROPHIC IMPACTS OF THE PERMIAN-TRIASSIC MASS EXTINCTION.

The Permian - Triassic Mass Extinction

CHAPTER FOUR

# THE PERMIAN-TRIASSIC EXTINCTION

### *Around 252 Million Years Ago*

The Permian-Triassic extinction, occurring approximately 252 million years ago, represents the most devastating of the five major mass extinction events, sometimes referred to as "The Great Dying." This extinction event is unparalleled in its scope and severity, eradicating around 90-96% of all marine species and roughly 70% of terrestrial vertebrate species. It reshaped the course of life on Earth, marking the end of the Paleozoic Era and the beginning of the Mesozoic Era, where the dinosaurs would eventually rise to prominence. The magnitude and mystery of the Permian-Triassic extinction make it one of the most studied and theorized events in Earth's history. This chapter delves into the causes, effects, and lasting impact of the extinction, providing a comprehensive analysis of what transpired during this critical juncture in the history of life.

**Setting the Stage: The Permian World**

To fully understand the impact of the Permian-Triassic extinction, it's essential to grasp the biological and geological context of the Permian period. Lasting from about 299 to 252 million years ago, the Permian period was a time of extraordinary geological and biological transformation. The supercontinent Pangaea had come together by this time, dramatically altering the global climate and creating extreme environments both on land and in the seas. A single landmass spanned nearly the entire Earth, leading to vast arid regions in the interior and coastlines exposed to extreme climate variations.

Life during the Permian period was diverse, with ecosystems flourishing both in the oceans and on land. Marine environments were teeming with a wide array of organisms, including brachiopods, ammonites, corals, and sponges. Trilobites, which had flourished for hundreds of millions of years, were nearing the end of their dominance, while crinoids, bryozoans, and various mollusks played essential roles in marine ecosystems. The oceans were also home to many primitive fish species and the ancestors of modern sharks.

On land, the early ancestors of mammals, known as synapsids, were among the dominant vertebrates, including species like *Dimetrodon* and *Edaphosaurus*. The Permian landscape was also populated by diverse reptilian groups, many of which would survive the extinction and evolve into dinosaurs during the Triassic. Plants such as conifers, ferns, and seed ferns (pteridosperms) flourished across the planet, forming vast forests that would contribute to the coal deposits still found today.

However, the late Permian was not without its challenges. Life during this period had already been subject to significant environmental changes, including fluctuating sea levels and shifting climates. Despite these stressors, ecosystems remained diverse and resilient. That resilience, however, would not be enough to withstand the cataclysmic events that were about to unfold.

**Causes of the Permian-Triassic Extinction**

The exact causes of the Permian-Triassic extinction are still the subject of scientific debate, and like many mass extinction events, the "Great Dying" likely resulted from a combination of factors. The extinction did not occur instantaneously but unfolded over hundreds of thousands to a few million years, with evidence pointing to a series of environmental stresses that compounded over time. The two leading hypotheses for the extinction's primary drivers are volcanic activity, particularly the Siberian Traps, and the subsequent environmental changes caused by climate shifts, ocean anoxia, and greenhouse gas emissions.

**Siberian Traps Volcanism: The Smoking Gun:** One of the most widely accepted hypotheses for the cause of the Permian-Triassic extinction is the massive volcanic eruptions of the Siberian Traps. Located in what is now Siberia, this large igneous province experienced one of the most extensive volcanic episodes in Earth's history, releasing vast quantities of lava, gas, and ash into the atmosphere. These volcanic eruptions, which occurred over a period of around one million years, covered an area of about 7 million square kilometers with layers of basalt, leading to some of the thickest and most expansive lava flows ever recorded.

The volcanic activity released massive amounts of carbon dioxide (CO2) and methane (CH4), both potent greenhouse gases, into the atmosphere. These gases contributed to a rapid and extreme greenhouse effect, causing a dramatic rise in global temperatures. Some estimates suggest that temperatures may have increased by as much as 8°C, creating inhospitable conditions for many species. The release of sulfur dioxide (SO2) from the volcanic eruptions likely caused acid rain, further damaging ecosystems by acidifying soils and bodies of water.

The gases emitted by the Siberian Traps volcanism not only had immediate effects on climate and the environment, but they also triggered longer-term consequences that compounded the devastation of life on Earth. For instance, the release of carbon dioxide would have significantly altered the carbon cycle, leading to more widespread ocean acidification, a factor that would have devastating impacts on marine life.

**Ocean Anoxia and Acidification: The Marine Apocalypse:** One of the most devastating consequences of the Permian-Triassic extinction was its impact on marine life, which bore the brunt of the extinction. A critical factor in the mass marine die-off was the development of widespread anoxia, or oxygen depletion, in the world's oceans. The rapid rise in global temperatures and the disruption of ocean circulation, driven by both volcanic activity and the altered climate, led to the stagnation of ocean water, particularly in the deep sea.

Under normal circumstances, ocean currents cycle oxygen-rich surface water to deeper parts of the ocean, maintaining a balance that supports marine ecosystems. However, during the Permian extinction, the disruption of these currents led to a catastrophic collapse of marine ecosystems. Vast regions of the ocean became devoid of oxygen, rendering them uninhabitable for most marine organisms. This is supported by evidence from the geological record, which shows the deposition of black shales during the late Permian. Black shales are sedimentary rocks that form in anoxic environments and contain high levels of organic material from dead organisms.

Ocean anoxia was further exacerbated by the release of methane from methane hydrates stored in ocean sediments. As temperatures increased, these hydrates were destabilized, releasing large amounts of methane into the oceans and atmosphere. Methane, being a far more potent greenhouse gas than carbon dioxide, contributed to even more rapid warming and increased the intensity of ocean anoxia. This feedback loop of warming, anoxia, and methane release created an environment in which marine life could not survive, leading to the extinction of many species, including the dominant reef-builders of the time.

Ocean acidification also played a significant role in the extinction of marine species. The massive influx of carbon dioxide from volcanic activity dissolved in seawater, forming carbonic acid, which lowered the pH of the oceans. Acidic conditions are particularly harmful to organisms that rely on calcium carbonate to build their shells and skeletons, such as corals, mollusks, and some plankton species. The collapse of coral reef ecosystems during the Permian extinction is one of the most striking examples of how ocean acidification can devastate marine life.

**Climate Change and Its Consequences for Terrestrial Life:** While the marine realm was hit hardest by the Permian-Triassic extinction, terrestrial ecosystems were also profoundly affected. The global climate underwent a dramatic transformation during this time, driven by the release of greenhouse gases from volcanic activity. As temperatures soared, the planet experienced widespread desertification, particularly in the interior regions of Pangaea, where the lack of moisture and extreme temperatures created inhospitable conditions for many terrestrial species.

The rapid warming of the climate led to the breakdown of ecosystems that had thrived during the Permian. Forests that had once dominated large areas of the supercontinent began to disappear as arid conditions spread, leading to the loss of habitats for many plant and animal species. The extinction of plant species, in turn, had cascading effects on herbivores and the predators that depended on them.

In addition to the rise in temperatures, the Permian-Triassic extinction was also marked by significant fluctuations in sea levels. As glaciers melted and thermal expansion caused the oceans to rise, many coastal ecosystems were inundated, while others were exposed as sea levels fell during periods of glaciation. These rapid and unpredictable changes in sea level further destabilized ecosystems, particularly in coastal areas where many species were already struggling to survive.

The extinction event also had a profound impact on terrestrial vertebrates, particularly the synapsids, the dominant land vertebrates of the Permian. Synapsids were the ancestors of mammals and had evolved a wide variety of forms, ranging from large, herbivorous species to smaller, carnivorous ones. The extinction event decimated synapsid populations, with many species disappearing entirely. However, a few synapsid groups managed to survive the extinction and would eventually give rise to the first true mammals in the Mesozoic.

**Methane Hydrate Release and Its Catastrophic Impact:** Another major contributor to the Permian-Triassic extinction was the potential release of methane hydrates, a type of methane gas trapped within ice crystals in ocean sediments. As global temperatures increased, these methane hydrates were destabilized, releasing vast quantities of methane into the atmosphere and oceans. Methane is a potent greenhouse gas, far more effective at trapping heat than carbon dioxide, and its release would have led to even more rapid and severe warming.

The release of methane hydrates may have created a positive feedback loop, where warming temperatures caused more methane to be released, which in turn caused even more warming. This feedback loop could have led to a "runaway greenhouse" effect, where the Earth's temperature rose uncontrollably, creating conditions that were hostile to most forms of life.

The release of methane would have also contributed to the development of ocean anoxia, as methane-consuming bacteria proliferated and used up the available oxygen in the water. This would have further exacerbated the oxygen depletion in the oceans, making it even more difficult for marine life to survive.

**The Impact on Life: A Catastrophic Loss of Biodiversity**

The Permian-Triassic extinction resulted in the largest loss of biodiversity in Earth's history. Entire ecosystems were obliterated, and the structure of life on Earth was forever altered. The extinction was particularly devastating to marine life, with around 90% of marine species wiped out. Among the groups that suffered the most were the trilobites, which had been a dominant group of marine arthropods for hundreds of millions of years. The Permian extinction marked the final extinction of the trilobites, leaving no descendants.

Other marine groups, such as brachiopods, ammonites, and crinoids, also suffered significant losses, although some managed to survive and would later diversify in the aftermath of the extinction. Reef-building organisms were particularly hard hit, with the collapse of coral reefs leading to the loss of many species that depended on these ecosystems for survival.

On land, the extinction led to the disappearance of many large synapsids, which had been the dominant terrestrial vertebrates of the Permian. However, a few synapsid groups, such as the cynodonts, managed to survive and would eventually give rise to the first mammals. The extinction also created new opportunities for other groups, such as the archosaurs, which would later evolve into the dinosaurs that dominated the Mesozoic.

The Permian-Triassic extinction event effectively wiped the slate clean, clearing the way for the evolution of new forms of life in the Triassic period. While the loss of biodiversity was immense, the extinction also set the stage for the rise of the dinosaurs and the eventual dominance of mammals in the Cenozoic Era.

**The Aftermath: Recovery and the Rise of New Life**

The recovery of life after the Permian-Triassic extinction was slow and uneven. It took millions of years for ecosystems to stabilize and for biodiversity to recover to pre-extinction levels. The early Triassic period, immediately following the extinction, was characterized by low levels of biodiversity and a harsh, unpredictable environment.

The oceans remained largely devoid of life for several million years, with only a few species of hardy survivors managing to eke out a living. Reef ecosystems, which had been devastated by the extinction, took particularly long to recover. It was not until the middle Triassic, around 10 million years after the extinction, that reefs began to re-establish themselves, and even then, they never regained the same level of complexity and diversity as they had during the Permian.

On land, the recovery was somewhat quicker, with new groups of reptiles and synapsids evolving to fill the ecological niches left vacant by the extinction. The archosaurs, which included the ancestors of dinosaurs and crocodiles, began to rise to prominence during the early Triassic, setting the stage for the dominance of the dinosaurs in the Mesozoic.

The Permian-Triassic extinction serves as a stark reminder of the fragility of life on Earth and the potential for environmental changes to cause catastrophic disruptions. It also highlights the resilience of life, as species that survived the extinction were able to adapt and evolve in response to the new challenges they faced. The extinction event shaped the future trajectory of life on Earth, leading to the rise of new groups of organisms and the eventual emergence of mammals, including humans.

**Conclusion: Lessons from the Permian-Triassic Extinction**

The Permian-Triassic extinction was the most severe extinction event in Earth's history, wiping out the majority of species and dramatically altering the course of evolution. Its causes were complex and likely involved a combination of factors, including volcanic activity, climate change, ocean anoxia, and methane hydrate release. The extinction event reshaped ecosystems both on land and in the oceans, leading to the loss of many dominant groups of organisms and the rise of new forms of life in the aftermath.

The lessons of the Permian-Triassic extinction are especially relevant today, as the modern world faces its own environmental challenges. The factors that contributed to the Great Dying, such as climate change and ocean acidification, are similar to the issues we face today as a result of human activities. Studying past extinction events can provide valuable insights into the potential consequences of environmental changes and the resilience of life in the face of adversity.

While the Permian-Triassic extinction was a catastrophe for life on Earth, it also set the stage for the evolution of new forms of life that would go on to dominate the planet for millions of years. It serves as a powerful reminder of the dynamic nature of life on Earth and the ability of species to adapt and evolve in response to even the most extreme challenges.

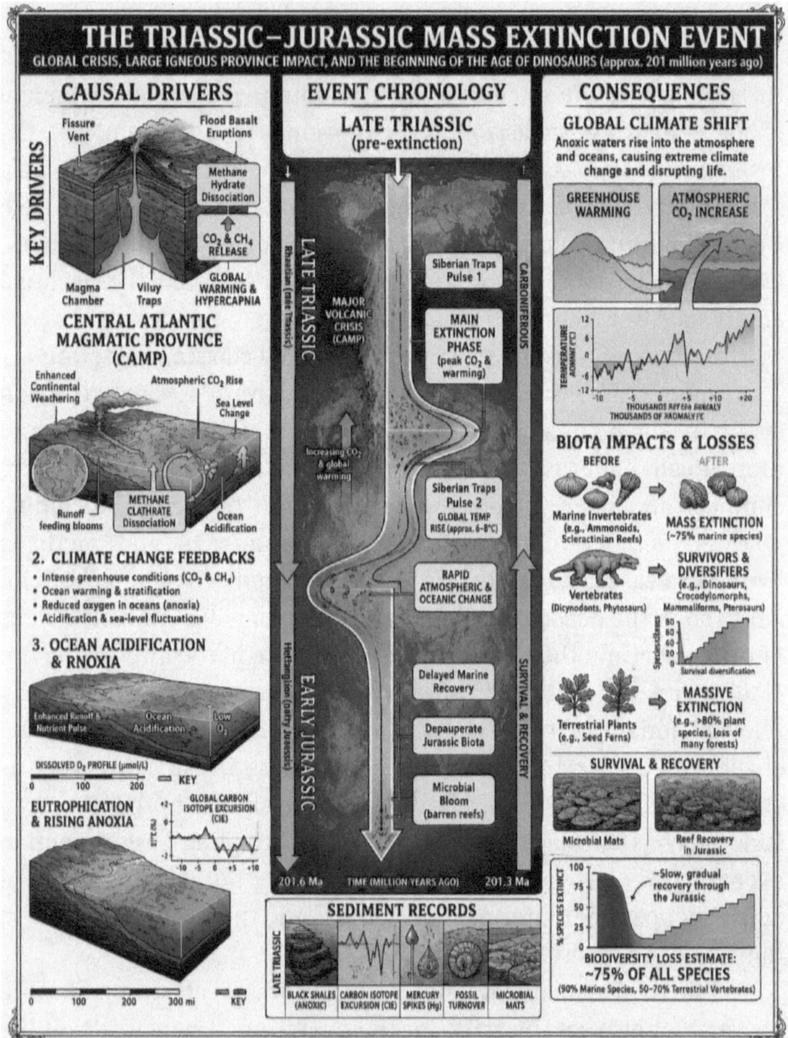


FIGURE 4.1: CAUSES, CHRONOLOGY, AND CATASTROPHIC IMPACTS OF THE TRIASSIC-JURASSIC MASS EXTINCTION.

The Triassic - Jurasic Mass Extinction

CHAPTER FIVE

# The Triassic-Jurassic Extinction

## *Around 201 Million Years Ago*

The Triassic-Jurassic extinction, occurring approximately 201 million years ago, marks one of the lesser-known but highly significant mass extinctions in Earth's history. This event occurred during the transition between the Triassic and Jurassic periods and reshaped life on the planet, particularly on land. While less severe than the Permian-Triassic extinction, the Triassic-Jurassic extinction wiped out around 70-75% of all species, clearing the way for the rise of dinosaurs as the dominant terrestrial vertebrates during the Mesozoic Era. The extinction had far-reaching effects on ecosystems both on land and in the oceans, with particular groups of animals, especially large amphibians and many reptiles, disappearing from the fossil record.

This chapter explores the complex causes behind the Triassic-Jurassic extinction, the catastrophic consequences for life on Earth, and how it paved the way for the dominance of dinosaurs. By understanding the environmental and biological upheavals during this period, we can gain a more comprehensive perspective on how mass extinctions shape the evolutionary path of life on Earth.

**The Triassic World: A Planet of Transition**

The Triassic period, which lasted from around 252 to 201 million years ago, was a time of recovery and reorganization following the devastation of the Permian-Triassic extinction, the largest mass extinction event in Earth's history. The early Triassic was characterized by a global ecosystem in recovery, with biodiversity levels slowly increasing after the near-total collapse of many marine and terrestrial environments.

During the Triassic, the supercontinent Pangaea was still largely intact, covering almost all the landmass on Earth. This vast continent contributed to a variety of extreme environmental conditions, from hot and arid deserts in its interior regions to lush coastal environments where life began to re-establish itself. The Triassic climate was generally hot and dry, with no evidence of polar ice caps, and seasonal variations in temperature would have been quite extreme, contributing to a highly dynamic and challenging environment for both plants and animals.

Life in the Triassic was diverse and thriving by the end of the period, particularly on land. The survivors of the Permian extinction were evolving and diversifying, with reptiles becoming increasingly dominant. Early forms of dinosaurs had already appeared, but they were relatively small and far from the dominant group that they would become in the Jurassic. Other groups of reptiles, such as archosaurs, which included crocodile-like creatures, and large amphibians, were among the dominant vertebrates on land. The synapsids, the ancestors of mammals, were still present but were no longer the dominant group of land vertebrates they had been in the Permian.

Marine life was also diversifying, with new species of fish, mollusks, and marine reptiles such as ichthyosaurs and plesiosaurs populating the oceans. Coral reefs, which had been devastated during the Permian extinction, were beginning to recover, and ammonites, a group of marine invertebrates related to modern cephalopods, were becoming more common.

The late Triassic saw the emergence of ecosystems that would characterize the Jurassic period, but it was a time of significant instability. By the end of the Triassic, life had rebounded from the Permian extinction, but it would soon face another environmental crisis that would again reshape the trajectory of life on Earth.

**Causes of the Triassic-Jurassic Extinction**

As with many mass extinction events, the Triassic-Jurassic extinction was likely caused by a combination of factors rather than a single catastrophic event. The exact causes are still debated, but the prevailing theories point to significant volcanic activity, climate change, ocean acidification, and the impact of changing atmospheric composition as the key drivers of this extinction. Some scientists have also speculated about the possible role of asteroid impacts, but there is less direct evidence for this hypothesis compared to other extinction events like the Cretaceous-Paleogene extinction.

**Volcanism: The Central Atlantic Magmatic Province (CAMP):** One of the most widely supported theories for the Triassic-Jurassic extinction is the massive volcanic activity associated with the breakup of the supercontinent Pangaea. During this period, Pangaea began to rift apart, eventually leading to the formation of the Atlantic Ocean. This tectonic activity triggered a series of enormous volcanic eruptions in a region known as the Central Atlantic Magmatic Province (CAMP). These eruptions covered vast areas of what is now eastern North America, northern Africa, and western Europe in thick layers of basalt, forming one of the largest igneous provinces in Earth's history.

The volcanic eruptions of CAMP released immense quantities of gases into the atmosphere, including carbon dioxide (CO2), sulfur dioxide (SO2), and methane (CH4). These gases had a profound impact on the climate, contributing to a runaway greenhouse effect that caused global temperatures to rise rapidly. The release of sulfur dioxide would have led to short-term cooling by reflecting sunlight away from the Earth's surface, but this cooling was followed by a longer-term warming due to the buildup of carbon dioxide and methane in the atmosphere.

The volcanic activity also contributed to the acidification of both the atmosphere and the oceans. Sulfur dioxide, when combined with water vapor in the atmosphere, forms sulfuric acid, which would have fallen as acid rain, damaging plant life and ecosystems on land. The release of carbon dioxide into the oceans caused ocean acidification, which had devastating effects on marine organisms, particularly those that relied on calcium carbonate to build their shells and skeletons.

**Climate Change and the Greenhouse Effect:** The volcanic eruptions of CAMP set off a chain reaction of environmental changes that severely disrupted the global climate. The release of large amounts of carbon dioxide into the atmosphere led to a significant greenhouse effect, where heat was trapped in the Earth's atmosphere, causing global temperatures to rise. This rapid warming would have caused severe stress on ecosystems, particularly those that were already vulnerable due to the instability of the late Triassic environment.

The global climate during the Triassic-Jurassic extinction likely became highly erratic, with extreme temperature fluctuations and disruptions to the water cycle. Evidence from the fossil record indicates that many species of plants and animals were unable to adapt to these rapid environmental changes. The hot, dry conditions that resulted from the greenhouse effect would have been particularly devastating for terrestrial ecosystems, leading to widespread desertification and the collapse of food chains.

The warming climate also had a profound impact on marine ecosystems. As ocean temperatures increased, the oxygen levels in the oceans decreased, leading to widespread anoxia, or oxygen depletion, in marine environments. This would have made it difficult for many marine species to survive, particularly those that lived in deep-sea environments. Ocean anoxia is one of the leading causes of marine extinctions during this period, as it created uninhabitable "dead zones" where life could not thrive.

**Ocean Acidification and Its Effects on Marine Life:** Ocean acidification played a crucial role in the extinction of marine species during the Triassic-Jurassic extinction. As volcanic activity released carbon dioxide into the atmosphere, much of this gas was absorbed by the oceans, where it dissolved to form carbonic acid. This process lowered the pH of seawater, making it more acidic. Acidic conditions are particularly harmful to marine organisms that rely on calcium carbonate to build their shells and skeletons, such as corals, mollusks, and some species of plankton.

The acidification of the oceans would have severely disrupted marine ecosystems, particularly coral reefs, which provide essential habitats for many marine species. The collapse of coral reefs during the Triassic-Jurassic extinction is well-documented in the fossil record, and it mirrors similar collapses during other mass extinctions, such as the Permian-Triassic extinction and the Cretaceous-Paleogene extinction. The loss of coral reefs would have had cascading effects on marine food webs, leading to the extinction of many species that depended on these ecosystems for survival.

In addition to the direct effects of acidification, the warming of the oceans would have exacerbated the problem by reducing the solubility of oxygen in seawater. As ocean temperatures rose, less oxygen was able to dissolve in the water, leading to hypoxic (low-oxygen) conditions that further stressed marine organisms. This combination of acidification, warming, and oxygen depletion created a hostile environment for marine life, contributing to the extinction of many species.

**The Role of Asteroid Impacts: A Hypothesis in Flux:** While asteroid impacts are the leading hypothesis for the Cretaceous-Paleogene extinction, their role in the Triassic-Jurassic extinction is less clear. Some scientists have suggested that asteroid impacts may have contributed to the extinction, either by triggering volcanic activity or by adding to the environmental stress already caused by other factors. However, there is currently no definitive evidence of a large impact crater dating to this time period, and most research focuses on the volcanic and climate-related causes of the extinction.

If asteroid impacts did play a role, it is likely that they acted as a secondary factor, exacerbating the environmental changes caused by volcanism and climate change. The possibility of multiple smaller impacts, rather than a single large impact, has also been suggested as a potential contributor to the extinction. However, until more evidence is uncovered, the role of asteroid impacts in the Triassic-Jurassic extinction remains speculative.

**The Impact on Life: A Changing World**

The Triassic-Jurassic extinction had profound effects on life both in the oceans and on land, leading to the disappearance of many species and the rise of new groups that would come to dominate the Jurassic period. The extinction event was particularly devastating for marine life, with a significant loss of biodiversity in the oceans. Many species of ammonites, bivalves, and marine reptiles disappeared, while coral reefs experienced a major collapse.

On land, the extinction event had a dramatic impact on vertebrate diversity, particularly among reptiles and amphibians. Large amphibians, such as the temnospondyls, were significantly affected, with many species going extinct. These amphibians had thrived during the late Triassic but were unable to adapt to the rapidly changing environmental conditions brought about by the extinction.

The reptiles of the Triassic period were diverse and included many groups that would later become dominant during the Jurassic. However, the extinction event led to a significant loss of diversity among these reptiles. The archosaurs, which included the ancestors of dinosaurs and crocodiles, were among the survivors, and they quickly expanded to fill the ecological niches left vacant by the extinction. This set the stage for the subsequent dominance of dinosaurs during the Jurassic period.

The Triassic-Jurassic extinction event also opened up new opportunities for the evolution of mammals. Although mammals had been present during the late Triassic, they were small and largely overshadowed by the dominant reptiles. The extinction of many reptilian groups allowed mammals to diversify and evolve into new forms, eventually leading to the rise of the mammals that would dominate the Cenozoic era.

**The Aftermath: Recovery and the Rise of Dinosaurs**

The recovery of life following the Triassic-Jurassic extinction was marked by a gradual resurgence of biodiversity, but the process took millions of years. In the aftermath of the extinction, ecosystems began to stabilize, and new groups of organisms emerged to fill the ecological roles left vacant by the extinctions. However, this recovery was not uniform, and some environments took longer to rebound than others.

In the oceans, it took time for coral reefs to recover from the devastation caused by the extinction. The loss of reef-building corals meant that the complex ecosystems associated with reefs were also lost. It was not until the early Jurassic that new coral species began to re-establish themselves, and even then, the complexity and diversity of reef ecosystems remained lower than in the Triassic. The recovery of marine life was also characterized by the emergence of new groups, including different types of bivalves and new forms of fish.

On land, the recovery was more pronounced, with the rise of dinosaurs becoming one of the most significant outcomes of the extinction. The ecological niches previously occupied by large reptiles and amphibians were rapidly filled by dinosaurs, which diversified into a wide range of forms, from small, agile theropods to massive sauropods. By the early Jurassic, dinosaurs had become the dominant terrestrial vertebrates, a status they would maintain for the remainder of the Mesozoic Era.

The Triassic-Jurassic extinction was not only a turning point for dinosaurs but also set the stage for the eventual emergence of mammals as a significant group. With the extinction of many large reptiles, mammals began to diversify and evolve into forms that would later give rise to the wide variety of mammals we see today. While mammals remained relatively small and inconspicuous during the Jurassic, the groundwork was laid for their subsequent rise to dominance in the Cenozoic era.

**Conclusion: The Legacy of the Triassic-Jurassic Extinction**

The Triassic-Jurassic extinction was a pivotal event in Earth's history, characterized by significant environmental changes that led to the loss of a substantial portion of biodiversity. Although not as severe as the Permian-Triassic extinction, it still had profound consequences for life on Earth, reshaping ecosystems both in the oceans and on land.

The causes of the extinction are complex and multifaceted, involving volcanic activity, climate change, ocean acidification, and possibly asteroid impacts. The interplay of these factors created a rapidly changing environment that many species were unable to adapt to, leading to a widespread collapse of ecosystems.

The aftermath of the extinction event ushered in a new era of life on Earth, with dinosaurs rising to dominate terrestrial ecosystems while new marine groups began to establish themselves in the oceans. The legacy of the Triassic-Jurassic extinction serves as a reminder of the fragility of ecosystems and the potential for rapid environmental changes to trigger significant shifts in biodiversity.

Understanding the events surrounding the Triassic-Jurassic extinction provides valuable insights into the processes of evolution, adaptation, and recovery following mass extinction events. It also highlights the ongoing challenges that modern ecosystems face in the face of human-induced climate change and environmental disruption, drawing parallels between past and present as we seek to understand the future of life on Earth.

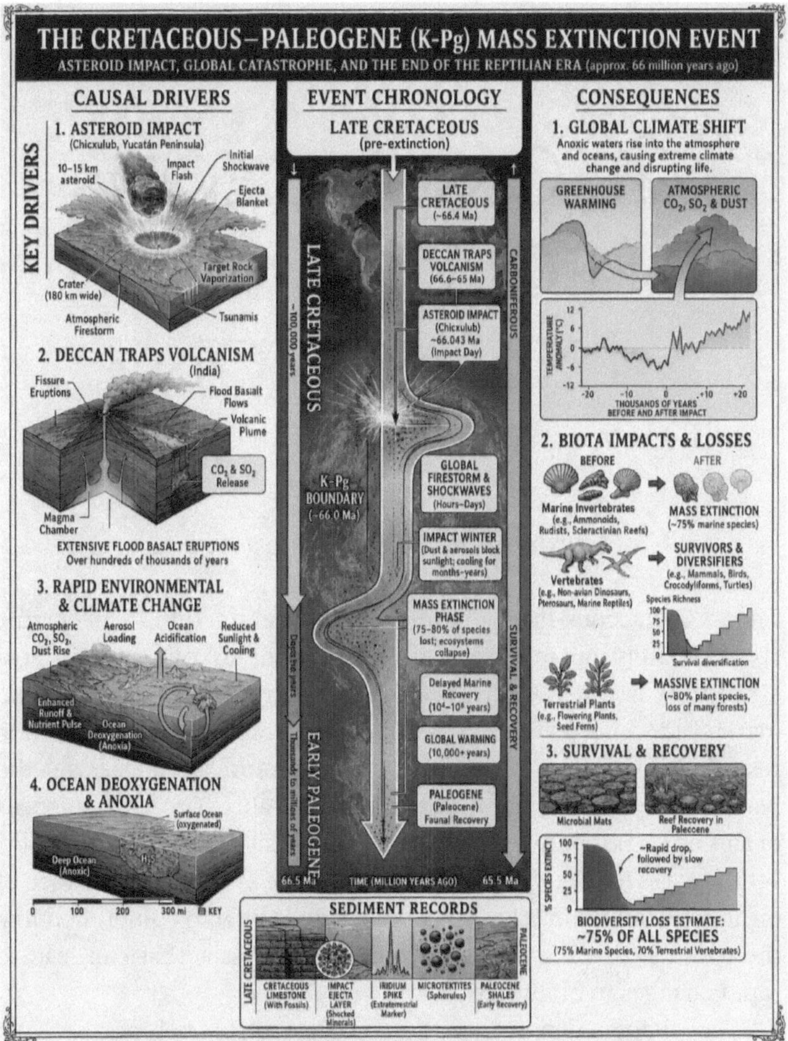


FIGURE 5.1: CAUSES, CHRONOLOGY, AND CATASTROPHIC IMPACTS OF THE CRETACEOUS–PALEOGENE (K-Pg) MASS EXTINCTION.

The Cretaceous - Paleogene (K-Pg) Mass Extinction

CHAPTER SIX

# The Cretaceous-Paleogene (K-Pg) Extinction

***Around 66 Million Years Ago***

The Cretaceous-Paleogene (K-Pg) extinction, occurring approximately 66 million years ago, stands as one of the most significant events in Earth's geological history. This mass extinction marked the end of the Cretaceous period and the Mesozoic Era, resulting in the loss of approximately 75% of all species on the planet, including the iconic non-avian dinosaurs. The consequences of this event not only reshaped the biodiversity of the Earth but also set the stage for the rise of mammals and the eventual dominance of birds. Understanding the K-Pg extinction is vital to comprehending the evolution of life on Earth and the processes that can lead to rapid changes in biodiversity.

This chapter delves into the complex causes behind the K-Pg extinction, examining the catastrophic impact of an asteroid or comet, the role of volcanic activity, and other environmental changes that contributed to the event. It will also explore the aftermath of the extinction, detailing how life on Earth recovered and evolved in the wake of such profound losses. Through a detailed examination of this mass extinction, we can appreciate the fragility of ecosystems and the resilience of life, drawing parallels to the challenges faced by contemporary organisms in a rapidly changing world.

**The Cretaceous World: A Time of Diversity**

The Cretaceous period, which lasted from approximately 145 to 66 million years ago, was a time of extraordinary diversity and flourishing ecosystems. During this period, dinosaurs dominated terrestrial environments, evolving into a wide range of forms and sizes. This era also witnessed the emergence of flowering plants, or angiosperms, which transformed terrestrial ecosystems and provided new sources of food for various species. Marine environments thrived as well, with a rich diversity of life, including mollusks, bony fish, and marine reptiles like mosasaurs and plesiosaurs.

The global climate during the Cretaceous was generally warm, with higher sea levels leading to extensive shallow inland seas. These seas provided rich habitats for marine life and contributed to the diversification of marine ecosystems. The Cretaceous is also notable for the presence of significant evolutionary developments, including the emergence of the first true mammals, which were small and often nocturnal creatures living in the shadows of the dominant reptiles.

Throughout the Cretaceous, ecosystems were characterized by intricate food webs and relationships among species. Dinosaurs occupied various ecological niches, from herbivores grazing on lush vegetation to formidable carnivores hunting prey. The evolution of flowering plants created new opportunities for insect pollinators, fostering co-evolution and further increasing biodiversity. However, this thriving world would soon face a catastrophic event that would drastically alter its trajectory.

**Causes of the K-Pg Extinction**

The causes of the K-Pg extinction have been extensively studied and debated among paleontologists and geologists. The prevailing theory attributes the mass extinction to a combination of catastrophic events, including an asteroid impact and extensive volcanic activity, both of which dramatically changed the Earth's climate and environment.

**The Asteroid Impact Hypothesis:** The most widely accepted explanation for the K-Pg extinction is the impact hypothesis, which suggests that a large asteroid or comet struck the Earth, leading to devastating environmental consequences. Evidence supporting this theory is found in the geological record, particularly in the form of a distinct layer of iridium-rich clay known as the K-Pg boundary layer. This layer is found in sedimentary rock strata around the world and marks the transition between the Cretaceous and Paleogene periods.

The discovery of the Chicxulub crater in the Yucatán Peninsula of Mexico provided compelling evidence for this hypothesis. Measuring approximately 150 kilometers (93 miles) in diameter, the impact site corresponds to the timing of the K-Pg extinction and suggests that a massive celestial object, estimated to be about 10 kilometers (6 miles) in diameter, struck the Earth with immense force. The impact would have released an energy equivalent to billions of atomic bombs, creating shockwaves and triggering massive fires across the planet.

The immediate effects of the impact would have been catastrophic. The energy released would have vaporized vast amounts of rock and released debris and dust into the atmosphere, blocking sunlight and leading to a dramatic drop in temperatures. This phenomenon, often referred to as "impact winter," would have severely disrupted photosynthesis, leading to a collapse of food chains both on land and in the oceans.

**Volcanism and Climate Change:** In addition to the impact hypothesis, another significant factor contributing to the K-Pg extinction was extensive volcanic activity in the Deccan Traps region of present-day India. The Deccan Traps represent one of the largest volcanic provinces on Earth, consisting of thick layers of basaltic lava flows that were created during a period of intense volcanic activity around the time of the K-Pg extinction.

The volcanic eruptions would have released vast amounts of gases, including sulfur dioxide and carbon dioxide, into the atmosphere. Sulfur dioxide would have contributed to short-term global cooling by forming sulfuric acid droplets in the atmosphere, while the release of carbon dioxide would have caused long-term warming. The combination of these effects would have led to significant climate instability, creating conditions that were inhospitable to many species.

The climatic changes caused by volcanic activity would have exacerbated the impacts of the asteroid strike. The release of volcanic gases and ash into the atmosphere would have further blocked sunlight, leading to reduced temperatures and altered weather patterns. This climatic disruption would have had a profound impact on ecosystems, affecting plant growth and the availability of food for herbivores and, in turn, for carnivores.

**The Role of Sea Level Changes and Ecosystem Disruption:** In addition to the asteroid impact and volcanic activity, changes in sea level during the late Cretaceous may have contributed to the extinction event. Fluctuations in sea level can significantly impact marine environments, leading to the loss of habitat for various species. During the late Cretaceous, the rising and falling of sea levels could have disrupted coastal and shallow marine ecosystems, further stressing marine life.

As sea levels rose, areas of land became submerged, leading to the loss of habitats for terrestrial organisms. Conversely, when sea levels dropped, coastal ecosystems were exposed, which could have caused additional stress on the flora and fauna adapted to these environments. These changes in habitat availability, combined with the catastrophic effects of the asteroid impact and volcanic activity, likely created a "perfect storm" that decimated biodiversity across the planet.

**The Impact on Life: A Catastrophic Turn**

The K-Pg extinction had immediate and far-reaching effects on life on Earth. The combined effects of the asteroid impact and volcanic activity resulted in a dramatic loss of biodiversity across multiple ecosystems, with both marine and terrestrial organisms suffering catastrophic declines.

**Marine Life: A Devastated Oceanic World:** The oceans, which had been teeming with life during the Cretaceous, experienced one of the most severe impacts from the K-Pg extinction. The immediate aftermath of the asteroid impact and subsequent climate changes led to a collapse of marine food webs. Species of plankton, which serve as the foundation of the marine food chain, were severely affected. The loss of phytoplankton due to the reduction in sunlight and the ensuing drop in photosynthesis caused cascading effects throughout the oceanic ecosystem.

Among the most notable casualties were the ammonites, a group of cephalopods that had thrived in the oceans for millions of years. The extinction of ammonites, along with many bivalves and other marine invertebrates, marked a significant loss of biodiversity in marine environments. Additionally, marine reptiles like mosasaurs and plesiosaurs, which were apex predators, also vanished from the fossil record.

The K-Pg extinction also had profound effects on the coral reefs, which experienced massive declines. Coral reefs are highly sensitive to environmental changes, and the combination of ocean acidification, increased water temperatures, and reduced sunlight would have devastated these complex ecosystems. The collapse of coral reefs had cascading effects on countless marine species that depended on these habitats for shelter and food.

**Terrestrial Life: Dinosaurs and Beyond:** The terrestrial ecosystems also faced drastic changes due to the K-Pg extinction. The most iconic victims of this extinction were, of course, the non-avian dinosaurs. These magnificent creatures had dominated terrestrial ecosystems for over 160 million years, but the abrupt environmental changes proved too much for many species to withstand. The loss of vegetation, coupled with a decline in available food sources, led to widespread extinctions among dinosaur populations.

While some smaller theropods are believed to have survived and eventually evolved into modern birds, the majority of the large dinosaur species disappeared. The extinction of dinosaurs opened up ecological niches that had been previously occupied by these massive reptiles, allowing for the diversification of other groups, particularly mammals.

In addition to dinosaurs, other terrestrial reptiles, including many groups of lizards and small mammals, were also affected by the extinction. The loss of biodiversity in terrestrial ecosystems meant that many niches became vacant, leading to a rapid evolution and diversification of the surviving species.

**The Extinction of Plants and Ecosystem Collapse:** The K-Pg extinction was not limited to animals; plant life also suffered significant losses. The abrupt changes in climate and the disruption of ecosystems led to a decline in many species of plants. The impact of the asteroid likely caused wildfires that destroyed vast swaths of vegetation, while the subsequent cooling and changing climate further stressed plant populations.

The loss of plant life had cascading effects on herbivorous animals, which relied on vegetation for sustenance. As food sources dwindled, many herbivorous species faced extinction, further impacting the carnivores that preyed on them. The decline of plant life also meant that many ecosystems were severely altered, leading to changes in soil composition and nutrient availability.

**The Aftermath: Recovery and the Rise of Mammals**

The recovery from the K-Pg extinction took millions of years and was characterized by a gradual resurgence of biodiversity. Following the mass extinction, ecosystems began to stabilize, and new groups of organisms emerged to fill the ecological roles left vacant by the extinctions. However, this recovery was not uniform, and some environments took longer to rebound than others.

**Marine Recovery: The Rise of New Groups:** In the oceans, recovery from the K-Pg extinction began relatively quickly, although it took millions of years for marine ecosystems to fully re-establish themselves. The extinction of many dominant marine species created opportunities for new groups to diversify and flourish. For example, the recovery of coral reefs took time, but eventually, new coral species began to re-establish themselves.

New groups of fish, including the rise of bony fish and sharks, began to diversify in the wake of the extinction. The collapse of larger marine reptiles created new ecological niches that were filled by different types of fish, leading to increased diversity in marine ecosystems.

**Terrestrial Recovery: The Age of Mammals:** On land, the aftermath of the K-Pg extinction paved the way for the rise of mammals. With the extinction of non-avian dinosaurs, mammals began to diversify and evolve into new forms. Initially small and nocturnal, mammals took advantage of the vacant ecological niches left behind by the dinosaurs. Over millions of years, mammals evolved into a wide range of forms, adapting to various environments and lifestyles.

The Paleogene period, which followed the K-Pg extinction, witnessed significant evolutionary developments in mammals. This era saw the emergence of larger mammals, including early forms of primates, ungulates, and carnivorous mammals. The diversification of mammals during this time laid the groundwork for the eventual emergence of modern mammalian groups.

The K-Pg extinction marked a turning point in the history of life on Earth. While it brought about the end of the age of dinosaurs, it also opened up new possibilities for the evolution and diversification of mammals. The ecosystems that emerged in the aftermath of the extinction were characterized by greater diversity and complexity, setting the stage for the rise of mammals as the dominant terrestrial vertebrates.

**The Lessons of the K-Pg Extinction:** The K-Pg extinction serves as a powerful reminder of the fragility of ecosystems and the potential for rapid environmental changes to trigger significant shifts in biodiversity. The combination of catastrophic events, including asteroid impacts and volcanic activity, resulted in a profound loss of life and a dramatic reshaping of Earth's ecosystems.

The study of the K-Pg extinction not only enhances our understanding of past biodiversity but also offers valuable insights into contemporary challenges facing ecosystems today. As human activities continue to exert unprecedented pressure on the environment, the lessons learned from mass extinction events can inform conservation efforts and highlight the importance of preserving biodiversity.

**Conclusion: The End of an Era and the Beginning of Another**

The Cretaceous-Paleogene extinction was a watershed moment in Earth's history, characterized by the dramatic loss of biodiversity and the end of the reign of dinosaurs. The causes of this extinction event are multifaceted, involving both extraterrestrial impacts and terrestrial volcanic activity. The immediate effects on ecosystems were catastrophic, leading to widespread extinctions among marine and terrestrial species.

In the aftermath of the K-Pg extinction, life on Earth exhibited remarkable resilience, with ecosystems gradually recovering and evolving into new forms. The extinction of non-avian dinosaurs opened the door for the rise of mammals, leading to the diversification of life that characterizes the modern world.

Understanding the K-Pg extinction provides essential insights into the dynamics of ecosystems, the processes of evolution, and the importance of biodiversity. As we face unprecedented environmental changes today, the lessons of this mass extinction resonate more than ever, urging us to recognize the interconnectedness of all life on Earth and the need to protect our planet's precious ecosystems for future generations.

FIGURE 6.1: DRIVERS, CHRONOLOGY, AND CATASTROPHIC IMPACTS OF THE SIXTH MASS EXTINCTION (ANTHROPOCENE – PRESENT DAY).

The Sixth Mass Extinction

CHAPTER SEVEN

# THE SIXTH EXTINCTION

### *The Anthropocene Extinction*

The Sixth Extinction, commonly referred to as the Anthropocene extinction, represents a profound and alarming moment in Earth's history. Unlike the five previous mass extinctions, which were driven by natural phenomena such as asteroid impacts, volcanic eruptions, and climate changes, the current extinction event is primarily the result of human activities. As we have entered the Anthropocene—a term used to describe the current geological epoch marked by significant human impact on the Earth's geology and ecosystems—the consequences of our actions have become increasingly evident. The Sixth Extinction is characterized by an unprecedented rate of species loss, threatening the delicate balance of life on Earth and fundamentally altering ecosystems worldwide.

This chapter will delve into the causes and consequences of the Sixth Extinction, examining the myriad ways in which human activities are driving species to extinction at alarming rates. We will explore the ecological, social, and ethical implications of this crisis and discuss potential paths forward to mitigate its impacts. Understanding the Sixth Extinction is crucial not only for the preservation of biodiversity but also for the survival of humanity itself, as the fate of our species is inextricably linked to the health of the planet's ecosystems.

**The Anthropocene: A New Epoch**

The concept of the Anthropocene, as an epoch distinct from the Holocene, emphasizes the profound impact that human beings have had on the Earth. This term, popularized by scientists in the late 20$^{th}$ century, signifies a new era characterized by significant alterations to the planet's geology, climate, and biodiversity due to human activities. From industrialization to urbanization, agriculture to deforestation, and pollution to climate change, humans have transformed the Earth in ways that were previously unimaginable.

In the Anthropocene, our influence on natural systems has reached a level where it can be compared to geological forces. This shift is evidenced by significant alterations in carbon cycles, nutrient cycles, and the hydrological cycle, all driven by human actions. The consequences of these alterations are profound, leading to habitat destruction, climate change, pollution, and the introduction of invasive species—all factors that contribute to the ongoing biodiversity crisis.

One of the defining features of the Anthropocene is the accelerated rate of species extinction. The current extinction rate is estimated to be 100 to 1,000 times higher than the natural background rate, a consequence of habitat destruction, overexploitation, pollution, and climate change. It is estimated that more than one million species are currently at risk of extinction, a sobering statistic that highlights the urgency of the crisis we face.

**The Drivers of the Sixth Extinction**

The Sixth Extinction is driven by a combination of interrelated factors, each contributing to the rapid decline of biodiversity. The primary drivers include habitat loss, climate change, overexploitation, pollution, and the introduction of invasive species. Each of these factors interacts with one another, creating a complex web of challenges that threaten the survival of countless species.

**Habitat Loss: The Fragmentation of Ecosystems:** Habitat loss is perhaps the most significant driver of the Sixth Extinction. As human populations continue to grow and expand, natural habitats are increasingly transformed into agricultural land, urban areas, and industrial zones. Deforestation, wetland drainage, and the conversion of grasslands for agriculture have led to the fragmentation and destruction of ecosystems across the globe.

The loss of habitat not only reduces the available space for wildlife but also disrupts ecological processes that are vital for the survival of species. Fragmented habitats can isolate populations, making it difficult for species to find mates, access food, and migrate in response to environmental changes. This isolation can lead to reduced genetic diversity, making populations more vulnerable to disease, environmental changes, and extinction.

Moreover, habitat destruction often leads to a cascade of secondary effects that further threaten biodiversity. For example, the loss of forests can result in increased soil erosion, altered water cycles, and the disruption of carbon storage, all of which have far-reaching impacts on ecosystems and the services they provide.

**Climate Change: The Shifting Climate Landscape:** Climate change is another critical factor driving the Sixth Extinction. The burning of fossil fuels, deforestation, and other human activities have led to significant increases in greenhouse gas concentrations in the atmosphere, resulting in global warming. The effects of climate change are profound and multifaceted, impacting ecosystems and species in various ways.

Rising temperatures, changing precipitation patterns, and more frequent extreme weather events can disrupt the delicate balance of ecosystems. Species that are unable to adapt to changing conditions or migrate to more suitable habitats face the risk of extinction. For instance, many amphibian populations are particularly vulnerable to climate change, as they rely on specific temperature and moisture conditions for breeding and survival.

In addition to direct impacts on species, climate change can also exacerbate other threats to biodiversity. For example, rising temperatures can increase the frequency and intensity of wildfires, further destroying habitats and threatening species. Similarly, changing ocean temperatures and acidity levels pose significant risks to marine ecosystems, including coral reefs, which are already experiencing widespread bleaching events.

**Overexploitation: The Unsustainable Harvesting of Resources:** The overexploitation of natural resources is another significant driver of the Sixth Extinction. Unsustainable hunting, fishing, and harvesting practices have led to dramatic declines in many species. For example, overfishing has severely depleted fish populations in oceans worldwide, disrupting marine ecosystems and threatening the livelihoods of coastal communities.

Wildlife trade, both legal and illegal, poses a significant threat to many species. The demand for exotic pets, traditional medicine, and luxury goods has led to the decline of numerous species, pushing them toward extinction. Iconic animals such as elephants and rhinos are particularly vulnerable to poaching, driven by the demand for ivory and horns, respectively.

The consequences of overexploitation extend beyond individual species. The loss of keystone species—those that play a critical role in maintaining the structure of an ecosystem—can lead to cascading effects that disrupt entire ecosystems. For example, the decline of top predators can lead to overpopulation of herbivores, resulting in overgrazing and further habitat degradation.

**Pollution: The Poisoning of Ecosystems:** Pollution is another critical factor contributing to the Sixth Extinction. The introduction of pollutants into ecosystems, whether through agricultural runoff, industrial waste, or plastic debris, poses significant risks to biodiversity. Chemicals such as pesticides, heavy metals, and plastics can have toxic effects on wildlife, leading to declines in populations and even local extinctions.

Aquatic ecosystems are particularly vulnerable to pollution. Nutrient runoff from agriculture can lead to algal blooms, which deplete oxygen levels in water and create "dead zones" where few organisms can survive. Additionally, plastic pollution has become a pervasive problem in oceans, with millions of tons of plastic entering marine environments each year. Marine species ingest or become entangled in plastic debris, leading to injury and death.

The impact of pollution on biodiversity extends beyond individual species. Disruption of ecosystem processes, such as nutrient cycling and decomposition, can have cascading effects that undermine the resilience of ecosystems. Healthy ecosystems are crucial for providing essential services such as clean water, air, and soil fertility, all of which are vital for human survival.

**Invasive Species: The Disruption of Native Ecosystems:** The introduction of invasive species poses a significant threat to biodiversity, particularly in isolated ecosystems. Invasive species often outcompete native species for resources, leading to declines in populations and even local extinctions. The introduction of non-native plants, animals, and pathogens can disrupt established ecological relationships, altering the dynamics of ecosystems.

For example, the introduction of the brown tree snake to Guam led to the decline and extinction of several native bird species. Similarly, invasive plants can outcompete native flora, altering habitat structures and reducing the availability of food and shelter for native wildlife.

The impacts of invasive species are often exacerbated by other drivers of the Sixth Extinction. For instance, habitat destruction can create opportunities for invasive species to establish themselves in previously intact ecosystems. Climate change can further facilitate the spread of invasive species by altering habitat conditions and allowing them to expand their ranges.

**The Consequences of the Sixth Extinction**

The consequences of the Sixth Extinction are far-reaching and profound, affecting not only the species that face extinction but also the ecosystems and human societies that depend on them. The loss of biodiversity has significant ecological, economic, and social implications, highlighting the interconnectedness of life on Earth.

**Ecological Impacts: Disruption of Ecosystem Services:** The extinction of species leads to the disruption of ecosystem services, which are the benefits that humans derive from functioning ecosystems. Biodiverse ecosystems provide essential services such as pollination, nutrient cycling, water purification, and climate regulation. The loss of species can undermine these services, leading to negative consequences for both natural systems and human societies.

For example, the decline of pollinators such as bees and butterflies threatens global food production, as many crops rely on these insects for pollination. Similarly, the loss of wetlands due to habitat destruction and pollution reduces the ability of these ecosystems to filter water, leading to increased water quality issues for communities that rely on freshwater sources.

The cascading effects of biodiversity loss can lead to ecosystem collapse, where the intricate relationships among species unravel, resulting in degraded habitats that can no longer support life. The interconnectedness of ecosystems means that the loss of one species can have far-reaching implications, affecting many others and ultimately jeopardizing the stability of entire ecosystems.

**Economic Consequences: The Cost of Biodiversity Loss:** The Sixth Extinction also carries significant economic consequences. The decline in biodiversity and the loss of ecosystem services can have direct impacts on industries that rely on natural resources, such as agriculture, fisheries, and tourism. The economic cost of biodiversity loss is challenging to quantify, but it is clear that the impacts are profound and far-reaching.

For example, overfishing not only threatens fish populations but also impacts coastal communities that depend on fisheries for their livelihoods. The loss of biodiversity in marine ecosystems can result in decreased fish stocks, leading to economic instability for those who rely on fishing as a primary source of income.

Furthermore, the loss of biodiversity can reduce the resilience of ecosystems to environmental changes, making them more vulnerable to disturbances such as climate change and natural disasters. This vulnerability can lead to increased costs for disaster response and recovery, further straining economic resources.

**Social Impacts: The Human Toll of Extinction:** The consequences of the Sixth Extinction extend beyond ecological and economic impacts, affecting human societies on multiple levels. Indigenous and local communities that rely on biodiversity for their cultural practices, subsistence, and livelihoods are particularly vulnerable to the impacts of species loss. As species decline and ecosystems degrade, these communities face the loss of traditional knowledge, cultural heritage, and food security.

The loss of biodiversity also poses ethical dilemmas, raising questions about our responsibilities to protect the planet's species and ecosystems. As stewards of the Earth, humans have a moral obligation to ensure the survival of other species and to recognize the intrinsic value of biodiversity. The extinction of species represents not only a loss of life but also a loss of cultural and ecological heritage that cannot be replaced.

Moreover, the impacts of the Sixth Extinction are not distributed equally. Vulnerable populations, particularly in developing countries, may bear the brunt of the consequences due to their dependence on natural resources for their livelihoods. As biodiversity declines, these communities may face increased food insecurity, loss of income, and diminished access to essential resources, exacerbating existing social inequalities.

**Mitigating the Sixth Extinction: Paths Forward**

Addressing the Sixth Extinction requires urgent and concerted action at local, national, and global levels. Mitigating the impacts of species loss and protecting biodiversity will necessitate a multifaceted approach that addresses the drivers of extinction while promoting sustainable practices and conservation efforts.

**Conservation and Restoration: Protecting Biodiversity Hotspots:** One of the most critical strategies for mitigating the Sixth Extinction is the establishment of protected areas and conservation initiatives aimed at preserving biodiversity hotspots. These regions, characterized by high levels of endemic species and unique ecosystems, play a vital role in maintaining global biodiversity. Protecting these areas from habitat destruction and exploitation is essential for ensuring the survival of many species.

In addition to protecting existing habitats, restoration efforts can help recover degraded ecosystems. Reforestation, wetland restoration, and habitat rehabilitation can enhance biodiversity and restore essential ecosystem services. Successful restoration projects not only benefit wildlife but also provide opportunities for sustainable livelihoods and community engagement.

**Sustainable Practices: Balancing Human Needs and Biodiversity:** Sustainable resource management practices are essential for reducing the impacts of human activities on biodiversity. This includes promoting sustainable agriculture, responsible fishing practices, and reducing pollution through cleaner production methods. By integrating ecological principles into resource management, we can reduce the pressure on ecosystems and help maintain biodiversity.

Education and awareness-raising initiatives can also play a crucial role in promoting sustainable practices. Engaging communities in conservation efforts and providing them with the knowledge and tools to protect their local ecosystems can foster a sense of stewardship and responsibility for biodiversity.

**Addressing Climate Change: Mitigating Global Threats:** Tackling climate change is essential for mitigating the impacts of the Sixth Extinction. Reducing greenhouse gas emissions, transitioning to renewable energy sources, and promoting energy efficiency are critical steps toward addressing the root causes of climate change. International cooperation and commitment to climate agreements, such as the Paris Agreement, are vital for limiting global warming and protecting biodiversity.

Adaptation strategies that enhance the resilience of ecosystems to climate change can also help mitigate the impacts on biodiversity. This includes creating wildlife corridors to facilitate species movement in response to changing habitats, protecting critical habitats, and implementing management practices that support ecosystem health.

**Ethical Considerations: Recognizing the Value of Biodiversity:** Addressing the Sixth Extinction also involves ethical considerations regarding our relationship with the natural world. Recognizing the intrinsic value of biodiversity and our responsibility to protect it is essential for fostering a culture of conservation. This includes acknowledging the rights of nature and advocating for policies that prioritize ecological health alongside human development.

Engaging diverse stakeholders, including indigenous communities, scientists, policymakers, and the public, in conservation efforts is crucial for building consensus and fostering collaborative action. By integrating diverse perspectives and knowledge systems, we can develop more effective and inclusive strategies for biodiversity conservation.

**Conclusion: The Imperative for Action**

The Sixth Extinction represents one of the most significant challenges facing humanity and the planet today. The accelerated loss of biodiversity driven by human activities threatens not only countless species but also the ecosystems and services that underpin human survival. Understanding the drivers and consequences of the Anthropocene extinction is crucial for developing effective strategies to mitigate its impacts.

The urgency of the situation demands immediate and concerted action. Protecting biodiversity, promoting sustainable practices, and addressing climate change are all essential components of the response to the Sixth Extinction. As stewards of the Earth, we have a moral obligation to protect the planet's rich biodiversity for future generations.

The choices we make today will determine the future of life on Earth. By recognizing the interconnectedness of all living beings and the importance of biodiversity, we can work toward a more sustainable and equitable future. The Sixth Extinction is not just a crisis; it is an opportunity to rethink our relationship with nature and forge a path toward a more harmonious coexistence with the planet's diverse ecosystems.

CHAPTER EIGHT

# Lessons from Past Extinctions

The story of life on Earth is marked by dramatic events of extinction that have reshaped the planet's biological landscape. Throughout geological history, mass extinctions have served as both punctuation marks and turning points, clearing the way for new forms of life and altering the course of evolution. While these events often signal catastrophic consequences for the species involved, they also offer valuable insights into the resilience and adaptability of life. This chapter explores the lessons we can glean from past extinctions, with a focus on the underlying causes, ecological ramifications, and potential pathways for recovery. By understanding these lessons, we can better inform our actions in the face of the ongoing biodiversity crisis and the Anthropocene extinction.

**Understanding Mass Extinctions**

Mass extinctions are defined as events during which a significant percentage of the planet's biodiversity is lost in a relatively short period. The geological record reveals five major mass extinctions, including the Ordovician-Silurian, Late Devonian, Permian-Triassic, Triassic-Jurassic, and Cretaceous-Paleogene extinctions. Each of these events was characterized by unique ecological dynamics, but they also shared commonalities in their underlying causes and consequences.

Mass extinctions serve as a stark reminder of the fragility of life and the interconnectedness of ecosystems. The loss of a single species can have cascading effects, leading to further extinctions and altering the structure and function of entire ecosystems. However, these events also highlight the resilience of life; ecosystems have shown an ability to recover and diversify following mass extinction events, leading to the emergence of new species and ecological communities.

**Causes of Past Extinctions**

Examining the causes of past extinctions provides valuable lessons for understanding the current biodiversity crisis. Several common themes emerge from the analysis of historical extinction events, revealing how environmental changes, both natural and anthropogenic, can trigger widespread species loss.

**Climate Change: A Recurring Catalyst:** Climate change has played a significant role in many past mass extinctions. Fluctuations in temperature, changes in sea levels, and shifts in atmospheric conditions have all contributed to the destabilization of ecosystems. For instance, during the Permian-Triassic extinction, rapid climate changes, including global warming and ocean acidification, resulted in widespread habitat loss and the collapse of marine ecosystems. Similarly, the end of the Cretaceous period saw dramatic shifts in climate patterns due to volcanic activity and asteroid impacts, leading to the extinction of the dinosaurs.

The lessons learned from these events emphasize the need for proactive measures to mitigate climate change and its potential impacts on biodiversity. As the planet continues to warm due to human activities, understanding the historical relationship between climate change and species extinction is critical for developing effective strategies to protect vulnerable ecosystems.

**Habitat Destruction: The Loss of Ecosystems:** Habitat destruction has been a recurring theme in the history of mass extinctions. As ecosystems are fragmented or destroyed, the species that depend on them face increased vulnerability. The end-Permian extinction, for example, was preceded by extensive habitat degradation due to volcanic activity, leading to the loss of vital ecosystems. Similarly, the ongoing Anthropocene extinction is characterized by habitat loss caused by urbanization, agriculture, and industrial development.

The lesson here is clear: protecting and restoring habitats is essential for preserving biodiversity. Efforts to conserve natural habitats, establish protected areas, and promote sustainable land-use practices can help mitigate the impacts of habitat destruction and support the resilience of ecosystems.

**Overexploitation: Unsustainable Resource Use:** Overexploitation of natural resources has been a significant driver of extinction events throughout history. During the Late Devonian extinction, excessive fishing and hunting of certain species contributed to their decline. The current rate of species loss due to overexploitation—whether through unsustainable fishing practices, hunting, or habitat degradation—echoes these historical precedents.

The lesson from past extinctions underscores the importance of sustainable resource management. Developing regulations and practices that prioritize the conservation of species and ecosystems is crucial for ensuring the long-term viability of our natural resources. This includes promoting responsible consumption, reducing waste, and advocating for policies that support sustainable practices in agriculture, forestry, and fisheries.

**Invasive Species: Disruption of Native Ecosystems:** The introduction of invasive species has been a significant factor in many extinction events, often exacerbating the pressures faced by native species. The arrival of non-native species can lead to competition for resources, predation, and the spread of diseases. The impact of invasive species was notably evident during the Cretaceous-Paleogene extinction when the arrival of mammals and birds displaced many dinosaur species.

The current biodiversity crisis is marked by the proliferation of invasive species in ecosystems worldwide. The lesson from past extinctions highlights the need for vigilance in monitoring and managing invasive species to protect native biodiversity. Implementing measures to prevent the introduction of non-native species, as well as controlling and eradicating invasive populations, is essential for preserving the integrity of ecosystems.

**Ecological Ramifications of Extinctions**

The ecological ramifications of past extinctions provide insights into the interconnectedness of species and ecosystems. The loss of biodiversity can lead to profound changes in ecosystem structure and function, with cascading effects that impact both wildlife and human populations.

**Trophic Cascades: The Ripple Effects of Extinction:** The concept of trophic cascades illustrates how the loss of one species can reverberate throughout an ecosystem. Trophic cascades occur when changes at the top of the food chain influence the abundance and distribution of species at lower trophic levels. For example, the extinction of large predators can lead to an increase in herbivore populations, resulting in overgrazing and habitat degradation. This, in turn, can further threaten plant species and disrupt the entire ecosystem.

Historical examples, such as the decline of top predators during the Permian-Triassic extinction, demonstrate the importance of maintaining biodiversity at all trophic levels. The lessons learned from these events emphasize the need to protect keystone species—those that play a critical role in maintaining ecological balance—as their loss can lead to irreversible changes in ecosystems.

**Ecosystem Resilience: The Path to Recovery:** Despite the devastating impacts of mass extinctions, ecosystems have shown remarkable resilience and the capacity for recovery over time. Following past extinction events, new species have emerged to fill ecological niches, leading to the diversification of life forms and the establishment of new ecosystems. For instance, after the Permian-Triassic extinction, the subsequent Triassic period saw the rise of reptiles and the eventual dominance of dinosaurs.

The recovery process, however, can take millions of years and is contingent upon several factors, including the availability of suitable habitats and the presence of species capable of adapting to changing conditions. The lessons from historical recoveries highlight the importance of conserving biodiversity and protecting ecosystems to enhance resilience and facilitate recovery in the face of future challenges.

**The Role of Evolution: Adaptation and Diversification:** Mass extinctions have historically served as catalysts for evolutionary change, creating opportunities for new species to emerge and diversify. The vacant ecological niches left by extinct species provide a blank slate for evolutionary experimentation, leading to the development of new adaptations and life forms. The aftermath of the Cretaceous-Paleogene extinction, for example, allowed mammals to diversify and eventually dominate terrestrial ecosystems.

Understanding the mechanisms of evolution and adaptation is crucial for anticipating how species may respond to contemporary challenges, including climate change and habitat loss. The lessons learned from past extinctions underscore the importance of preserving genetic diversity, which is essential for species to adapt to changing environmental conditions.

**Ethical Considerations and Human Responsibility**

The lessons from past extinctions extend beyond ecological and evolutionary insights; they also raise profound ethical questions regarding our responsibility to the planet and its inhabitants. As stewards of the Earth, humans have a moral obligation to protect biodiversity and ensure the survival of other species.

**Recognizing Intrinsic Value:** The intrinsic value of biodiversity—the inherent worth of all living organisms—must be acknowledged in discussions of conservation and extinction. Each species plays a unique role in the web of life, contributing to the overall health and functioning of ecosystems. The loss of a single species can have far-reaching consequences, not only for other species but also for human societies that rely on biodiversity for their well-being.

Recognizing the intrinsic value of biodiversity compels us to take action to protect species and ecosystems, not merely for their utility to humans but for their own sake. This ethical perspective can guide conservation efforts and inspire a deeper appreciation for the natural world.

**Equity and Justice in Conservation:** The impacts of past and present extinctions are not felt equally across societies. Vulnerable communities, particularly in developing regions, often bear the brunt of biodiversity loss, facing increased food insecurity, loss of livelihoods, and diminished access to natural resources. The lessons from past extinctions highlight the importance of equity and justice in conservation efforts.

Efforts to protect biodiversity must consider the needs and rights of local communities, particularly those that rely on natural resources for their survival. Engaging these communities in conservation initiatives can foster a sense of ownership and responsibility, leading to more effective and sustainable outcomes.

**A Call to Action: Protecting Our Planet:** The urgency of the current biodiversity crisis demands immediate and coordinated action. The lessons learned from past extinctions underscore the need for a proactive approach to conservation, emphasizing the importance of protecting habitats, reducing resource exploitation, and addressing the drivers of extinction.

This call to action requires collaboration among governments, non-governmental organizations, scientists, and local communities to develop effective strategies for biodiversity conservation. Policies that prioritize sustainability, habitat protection, and community engagement are essential for mitigating the impacts of the ongoing Anthropocene extinction.

**Conclusion: Building a Resilient Future**

The lessons from past extinctions serve as a powerful reminder of the interconnectedness of life and the fragility of ecosystems. Understanding the causes and consequences of historical extinction events provides valuable insights that can inform our actions in the face of contemporary challenges.

As we confront the ongoing biodiversity crisis, the lessons from past extinctions compel us to recognize our role as stewards of the planet. By protecting biodiversity, promoting sustainable practices, and advocating for equitable conservation efforts, we can work toward a more resilient future for both human societies and the natural world.

Ultimately, the story of past extinctions is not one of despair but of hope and resilience. The capacity of life to adapt and recover in the face of adversity serves as a testament to the enduring power of nature. By learning from the past and taking action in the present, we can forge a path toward a more sustainable and equitable future, ensuring that the lessons of history guide our efforts to protect the planet's rich tapestry of life for generations to come.

CHAPTER NINE

# THE FUTURE OF EARTH'S BIODIVERSITY

As we stand at the dawn of the Anthropocene epoch, the future of Earth's biodiversity hangs in a delicate balance. Human activities are altering the natural world at an unprecedented pace, pushing many species toward extinction while simultaneously reshaping ecosystems and environmental conditions. Understanding the trajectories that biodiversity may take in the future requires not only a thorough examination of current trends and threats but also an exploration of the potential pathways toward recovery and sustainability. This chapter delves into the critical aspects of the future of biodiversity, exploring the challenges we face, the potential for resilience and adaptation, and the imperative for conservation and sustainable practices.

**The Current State of Biodiversity**

To appreciate the future of Earth's biodiversity, it is essential to grasp its current state. The Global Biodiversity Outlook (GBO) and other assessments provide a stark picture: species loss is accelerating, ecosystems are deteriorating, and the planet's natural capital is in decline. According to the Intergovernmental Science-Policy Platform on Biodiversity and Ecosystem Services (IPBES), up to one million species face extinction, many within the next few decades if current trends continue. This crisis is primarily driven by habitat destruction, climate change, pollution, overexploitation, and the introduction of invasive species.

The implications of biodiversity loss extend beyond the intrinsic value of species. Ecosystems provide vital services that sustain human life, including clean air and water, pollination of crops, soil fertility, and climate regulation. The decline of biodiversity jeopardizes these services, posing significant risks to food security, human health, and overall well-being. The future of biodiversity, therefore, is not only an ecological concern but a pressing human issue that requires immediate attention and action.

**Drivers of Change**

Understanding the key drivers of biodiversity loss is crucial for addressing the challenges that lie ahead. These drivers are complex and interconnected, and they interact in ways that exacerbate the overall crisis.

**Climate Change and Its Impacts:** Climate change is perhaps the most significant and far-reaching threat to biodiversity. Rising temperatures, shifting precipitation patterns, and increasing frequency of extreme weather events are altering habitats and ecosystems across the globe. Many species struggle to adapt to these rapid changes, leading to shifts in distribution, changes in reproductive cycles, and, in some cases, local extinctions.

The IPCC (Intergovernmental Panel on Climate Change) reports that if global temperatures rise by 1.5 degrees Celsius, approximately 20-30% of species assessed will be at an increased risk of extinction. Ecosystems such as coral reefs are particularly vulnerable; with rising ocean temperatures, coral bleaching events are becoming more frequent, threatening the myriad species that depend on these ecosystems for survival.

The future will require a multifaceted approach to climate change mitigation and adaptation, integrating conservation strategies that account for the impacts of changing climatic conditions on species and ecosystems.

**Habitat Destruction and Fragmentation:** The conversion of natural habitats for agriculture, urban development, and infrastructure continues to escalate, leading to the fragmentation and degradation of ecosystems. Deforestation, wetland drainage, and land-use change significantly impact the availability of habitat for many species. Fragmented habitats isolate populations, reducing genetic diversity and making them more susceptible to extinction.

In the future, it will be crucial to balance human development with conservation efforts. This may involve the establishment of wildlife corridors that connect fragmented habitats, allowing species to migrate, breed, and adapt to changing environments. Additionally, the promotion of sustainable land-use practices can help mitigate the impacts of habitat destruction.

**Pollution: The Silent Killer:** Pollution, whether through plastics, chemicals, or nutrient runoff, poses a significant threat to biodiversity. Water bodies, soils, and air quality suffer from pollution, leading to toxic environments that are inhospitable to many species. For instance, the excess use of fertilizers leads to eutrophication, creating dead zones in aquatic environments where few organisms can survive.

Addressing pollution requires a global commitment to reducing waste, improving waste management systems, and adopting sustainable agricultural practices. Public awareness and policy changes are essential to mitigate pollution's impact on ecosystems and protect biodiversity.

**Invasive Species and Disease:** The introduction of non-native species, often facilitated by global trade and travel, disrupts local ecosystems and outcompetes native species for resources. Invasive species can lead to significant declines in biodiversity, particularly when they are coupled with other stressors like climate change and habitat destruction.

Furthermore, the emergence of diseases, both in wildlife and in humans, poses an additional threat to biodiversity. For example, the chytrid fungus has devastated amphibian populations worldwide. As ecosystems are disrupted, the likelihood of disease outbreaks increases, further complicating conservation efforts.

The future must prioritize biosecurity measures to prevent the introduction of invasive species and monitor the health of ecosystems. Proactive strategies, including early detection and rapid response plans, are essential for safeguarding native biodiversity.

**The Path to Recovery and Sustainability**

Despite the daunting challenges that lie ahead, there is potential for recovery and resilience in ecosystems. Lessons from past extinctions and ongoing conservation efforts provide valuable insights into how biodiversity can be preserved and restored.

**Conservation Strategies: A Multifaceted Approach:** The future of biodiversity will depend on a multifaceted approach to conservation that incorporates a variety of strategies tailored to specific ecosystems and species. Protected areas are vital for conserving biodiversity, as they provide refuges for endangered species and critical habitats. Expanding and effectively managing protected areas can help safeguard biodiversity in the face of ongoing threats.

In addition to traditional protected areas, novel conservation approaches such as biodiversity offsets, habitat restoration, and rewilding can play a significant role in promoting resilience. Biodiversity offsets involve compensating for biodiversity loss in one area by restoring or protecting biodiversity in another. Similarly, habitat restoration aims to rehabilitate degraded ecosystems, enhancing their capacity to support diverse life forms.

**Community Engagement and Empowerment:** Successful conservation efforts must involve local communities in decision-making processes and empower them to take ownership of biodiversity initiatives. Indigenous and local knowledge is invaluable in understanding ecosystems and developing sustainable practices. By integrating traditional ecological knowledge with scientific research, conservation strategies can be more effective and culturally relevant.

Community-based conservation initiatives can lead to more sustainable resource management, enhance local livelihoods, and promote biodiversity conservation. Engaging communities fosters a sense of stewardship, encouraging individuals to protect their natural environments for future generations.

**Technological Innovations: Tools for Conservation:** The future of biodiversity conservation will increasingly rely on technological innovations. Advances in remote sensing, data analytics, and biotechnology offer new tools for monitoring ecosystems and species populations. For example, satellite imagery can be used to assess habitat changes, while genetic tools can aid in the identification and conservation of endangered species.

Citizen science initiatives, which involve the public in data collection and monitoring, can also enhance our understanding of biodiversity trends. Harnessing technology and public engagement can significantly improve conservation outcomes and contribute to more informed decision-making.

**Restoration Ecology: Healing the Earth:** Restoration ecology focuses on the rehabilitation of degraded ecosystems, seeking to restore their function and biodiversity. Efforts to restore habitats can lead to increased resilience against climate change and enhance the overall health of ecosystems. This may involve replanting native vegetation, removing invasive species, and restoring natural processes such as fire regimes.

Successful restoration projects not only benefit biodiversity but also provide essential ecosystem services, contributing to human well-being. The future of biodiversity will be inextricably linked to our ability to heal and restore the planet.

**Ethical Considerations and Global Responsibility**

The future of Earth's biodiversity raises significant ethical questions regarding our responsibility to the planet and its inhabitants. As the stewards of the Earth, humans have an obligation to protect and preserve the natural world for future generations.

**Intergenerational Equity:** The concept of intergenerational equity emphasizes the need to consider the rights and needs of future generations when making decisions that impact biodiversity. Our actions today will have profound implications for the health of ecosystems and the survival of species in the future.

Embracing this ethical framework can inspire individuals, communities, and governments to prioritize conservation efforts and sustainable practices. By safeguarding biodiversity today, we ensure that future generations inherit a healthy planet capable of sustaining life.

**Global Cooperation and Collaboration:** Biodiversity loss is a global issue that transcends national boundaries. The future of Earth's biodiversity necessitates international cooperation and collaboration. Agreements such as the Convention on Biological Diversity (CBD) provide frameworks for countries to work together in conserving biodiversity and sharing resources.

Global initiatives, such as the United Nations' Sustainable Development Goals (SDGs), emphasize the importance of biodiversity for achieving sustainability. By fostering collaboration among nations, organizations, and communities, we can collectively address the challenges facing biodiversity and work toward a more sustainable future.

**Conclusion: A Shared Vision for the Future**

The future of Earth's biodiversity is uncertain, marked by both profound challenges and remarkable opportunities. As we confront the ongoing biodiversity crisis, it is imperative to recognize the interconnectedness of all life on Earth and the vital role that biodiversity plays in sustaining ecosystems and human societies.

Learning from the past and embracing innovative conservation strategies will be essential for safeguarding biodiversity in the face of rapidly changing conditions. By prioritizing community engagement, leveraging technology, and committing to ethical stewardship, we can work towards a resilient future where biodiversity thrives.

Ultimately, the future of Earth's biodiversity is a shared responsibility. Every action taken today—whether through conservation efforts, sustainable practices, or public advocacy—contributes to a collective vision of a healthier planet. By recognizing the intrinsic value of biodiversity and committing to its protection, we can ensure that the Earth remains a vibrant tapestry of life for generations to come.

CHAPTER TEN

# CONCLUSION

## *A Fragile Symphony of Life and Loss*

As we reach the final pages of this journey through Earth's deep and turbulent biological history, we are left with a sobering yet deeply illuminating understanding of what it means to be part of this ever-evolving planet. The story of mass extinctions is not merely one of loss—it is a narrative of transformation, upheaval, resilience, and rebirth. Each chapter of Earth's past extinction events reveals not only how fragile life can be, but also how incredibly adaptive it is. However, in examining the Sixth Extinction—a crisis unfolding within our own lifetimes—we come face to face with the uncomfortable truth that human beings are no longer mere witnesses to extinction, but its primary drivers.

Throughout this book, we have traversed the epic timelines of extinction: the frozen silence of the Ordovician-Silurian period, the slow suffocation of marine ecosystems in the Devonian, the cataclysmic fires of the Permian-Triassic, the tectonic reshaping of the Triassic-Jurassic boundary, and the abrupt, fiery impact that ended the age of dinosaurs at the Cretaceous-Paleogene boundary. Each of these events tore through Earth's biosphere, eliminating species wholesale, emptying niches, silencing songs that had once echoed for millions of years. And yet, after each event, life rebounded—not instantly, and not always in the same form—but with an astounding creative force, reweaving the web of ecosystems in new and unpredictable patterns.

In the fossil record, we read these stories like detectives piecing together a mystery from fragmented clues: the sudden disappearance of trilobites, the proliferation of fern spikes after mass forest die-offs, the iridium layer marking the extraterrestrial blow that sealed the fate of non-avian dinosaurs. All of these clues speak to the same phenomenon—mass extinction is a rupture, a severing of continuity, but it is also a threshold. What lies beyond that threshold, however, depends on what survives, and what conditions shape the aftermath.

As we look toward the present and future, the Sixth Extinction challenges us to view ourselves not only as students of Earth's history, but as stewards of its future. Unlike previous mass extinctions, this one is being driven by a single species: Homo sapiens. With our capacity for technology, industry, and dominion over land, sea, and air, we have disrupted natural balances that evolved over millions of years. Our hunting, farming, fishing, mining, polluting, and warming of the planet have driven species to the brink or over the edge of extinction at a pace that outstrips even the most violent episodes in the geological record.

What makes the Sixth Extinction especially poignant is that we are aware of it. The dinosaurs did not have the capacity to ponder their fate; trilobites could not speculate on ecological disruption. We can. We have scientific institutions, satellites, ecological models, and genetic databases—tools that allow us to predict trends, assess risks, and imagine alternatives. And yet, the paradox of modern humanity is that with all this knowledge, we often remain inert, locked in patterns of consumption and growth that push ecosystems past tipping points.

The ultimate question posed by this moment in time is: what role will we play in the next chapter of life on Earth? Will we be a destructive force that collapses biodiversity and narrows the future of evolution? Or will we become guardians of planetary resilience, using our intelligence and compassion to reverse the tide?

To answer this, we must reflect on the lessons from past extinctions. Each one teaches something unique. The Ordovician-Silurian event shows how climatic shifts—such as glaciation—can swiftly reshape the oceanic and atmospheric dynamics of the planet. The Devonian extinction reveals how changes in plant life and nutrient cycles can trigger long-term ecological imbalances. The Permian-Triassic cataclysm is a cautionary tale of greenhouse gases and volcanic instability driving an ecosystem collapse on an unimaginable scale. The Triassic-Jurassic transition reminds us of how life responds to tectonic and climatic upheaval, often unpredictably. And the Cretaceous-Paleogene extinction teaches the power of sudden catastrophe to remake the rules of survival overnight.

We also learn that extinction is not just a moment, but a process. The Earth's biosphere does not snap back after a die-off; it takes millions of years for diversity to recover, and even longer for ecosystems to stabilize. This means that the actions we take today will echo for geological epochs. We are shaping not only our own legacy, but the evolutionary possibilities for all life to come.

Another important lesson lies in the idea of thresholds. Each extinction event occurred when planetary systems reached a breaking point. Ecosystems are flexible, adaptive, and complex, but they are not invincible. Once feedback loops are triggered—be it runaway warming, acidification, habitat collapse, or species interdependency failures—the cascade becomes unstoppable. The closer we push toward these thresholds in the modern day, the more we gamble with irreversible outcomes.

And yet, not all is doom. The concept of extinction is inseparable from the concept of renewal. It is a paradox that the darkest chapters in life's history also gave rise to bursts of innovation. Mammals emerged in the shadow of dinosaurs. Birds evolved from surviving theropods. Coral reefs reassembled, forests regrew, and oceans repopulated. But that recovery came at the cost of time and suffering—and of irretrievable loss. While life as a whole is resilient, individual species, especially those that are highly specialized or slow to reproduce, are not.

In the Anthropocene, our capacity to destroy is matched by our potential to protect. We are already seeing efforts that show what is possible when knowledge meets action: rewilding landscapes, restoring coral reefs, creating conservation corridors, and preserving genetic material through seed banks and biorepositories. Indigenous communities around the world demonstrate sustainable ways of living that respect ecological rhythms, offering models for how humanity might re-root itself in the web of life rather than standing above it.

Moreover, technological advances offer a double-edged sword. On one side, they accelerate destruction through deforestation, overfishing, and emissions. On the other, they offer tools for monitoring endangered species, mapping ecosystems, and reducing our ecological footprint. The challenge lies in how we choose to wield this power. Will our future technologies serve ecosystems, or continue to exploit them?

Ethically, the Sixth Extinction raises profound questions about responsibility. What do we owe to other species? Is it enough to avoid causing harm, or must we also actively restore what we have damaged? How do we weigh human development needs against the intrinsic value of biodiversity? These are not merely philosophical musings—they are decisions being made every day in policy rooms, boardrooms, and communities across the globe.

Culturally, we must reframe how we see ourselves in relation to nature. The myth of human exceptionalism—that we stand outside and above the natural world—has led us down a path of disconnection. But biology tells a different story: that we are one branch among many on the great tree of life, connected through shared genes, common ancestors, and mutual dependencies. Our fate is bound to the health of pollinators, fungi, forests, and oceans. When we destroy them, we undermine our own foundations.

Spiritually, many find in the story of extinction a call to humility. The Earth has existed for 4.5 billion years. Life has survived asteroid impacts, global ice ages, supervolcanoes, and the shifting of continents. Human civilization, by contrast, is only a few thousand years old. Recognizing our smallness in this grand timeline is not a defeat, but an invitation—to live more wisely, more gently, more attuned to the rhythms of the planet.

As we close this book, let us carry forward not only knowledge, but a sense of responsibility. Mass extinctions are not relics of the past; they are warnings etched in stone, coral, and bone. They are reminders that life is both fragile and fierce. That ecosystems are webs, not pyramids. That survival is not guaranteed by strength alone, but by adaptability, cooperation, and awareness.

We are at a crossroads. The Sixth Extinction is not yet complete. Its outcome is not written in fossils or fated by volcanoes. It is being written by human hands—through legislation, innovation, education, and action. Each species we save, each forest we protect, each lifestyle choice we make that honors the web of life, is a step away from collapse and toward continuity.

In honoring the memory of Earth's past extinctions, let us also become authors of a new story—one in which future generations inherit not a barren planet, but a thriving, diverse, and enduring Earth. That is the ultimate lesson. That is the legacy we have the power to shape.

# Afterword

As the final words of this book settle into silence, what remains is not just a story of death, but of awareness. The journey through Earth's mass extinctions is not merely an academic odyssey through geological eras and fossilized bones—it is a mirror held up to our time. It reflects the wonder, the peril, and the profound responsibility that now rests in human hands.

This book began with the intention of understanding the vast tapestry of life on Earth and how, at five separate points in time, that tapestry was torn apart by forces beyond imagination. Each extinction brought an end to millions of years of evolution, silencing entire ecosystems, vanquishing dominant species, and redrawing the blueprint of survival. In studying these events, we found more than just patterns of destruction; we discovered the enduring will of life to adapt, renew, and emerge in unfamiliar forms.

But this final mass extinction, the one we are now living through is different. It is not the slow hand of geological time or the random strike of a cosmic body that endangers life. It is us. It is our inventions, our industries, our appetites, and, above all, our indifference that have become agents of erasure. This knowledge is both devastating and empowering. Because if we are the cause, then we can also be the cure.

The Sixth Extinction is still unfolding. Species continue to vanish. Forests fall. Coral reefs bleach. But even now, hope flickers. Movements rise. Youth voices demand accountability. Indigenous knowledge reclaims its place. Conservationists, scientists, artists, and ordinary people act as planting, protecting, protesting, teaching, and healing.

In the vast story of Earth, humanity is a young chapter. Our destiny is not yet fossilized. What we do today will echo for millions of years. Future generations may look back at this moment as a turning point, either the beginning of irreversible decline or the dawn of planetary stewardship.

The afterword is not just the end of a book; it is the beginning of reflection. And perhaps, of action. Let this story not end with extinction, but with awakening. Let it stir something deeper than fear; let it awaken reverence. For life. For Earth. For all we stand to lose, and all we still have the power to protect.

The page turns. The story continues. And now, it is ours to write.